FRANK AUERBACH

London Building Sites 1952-62

THE COURTAULD Gallery

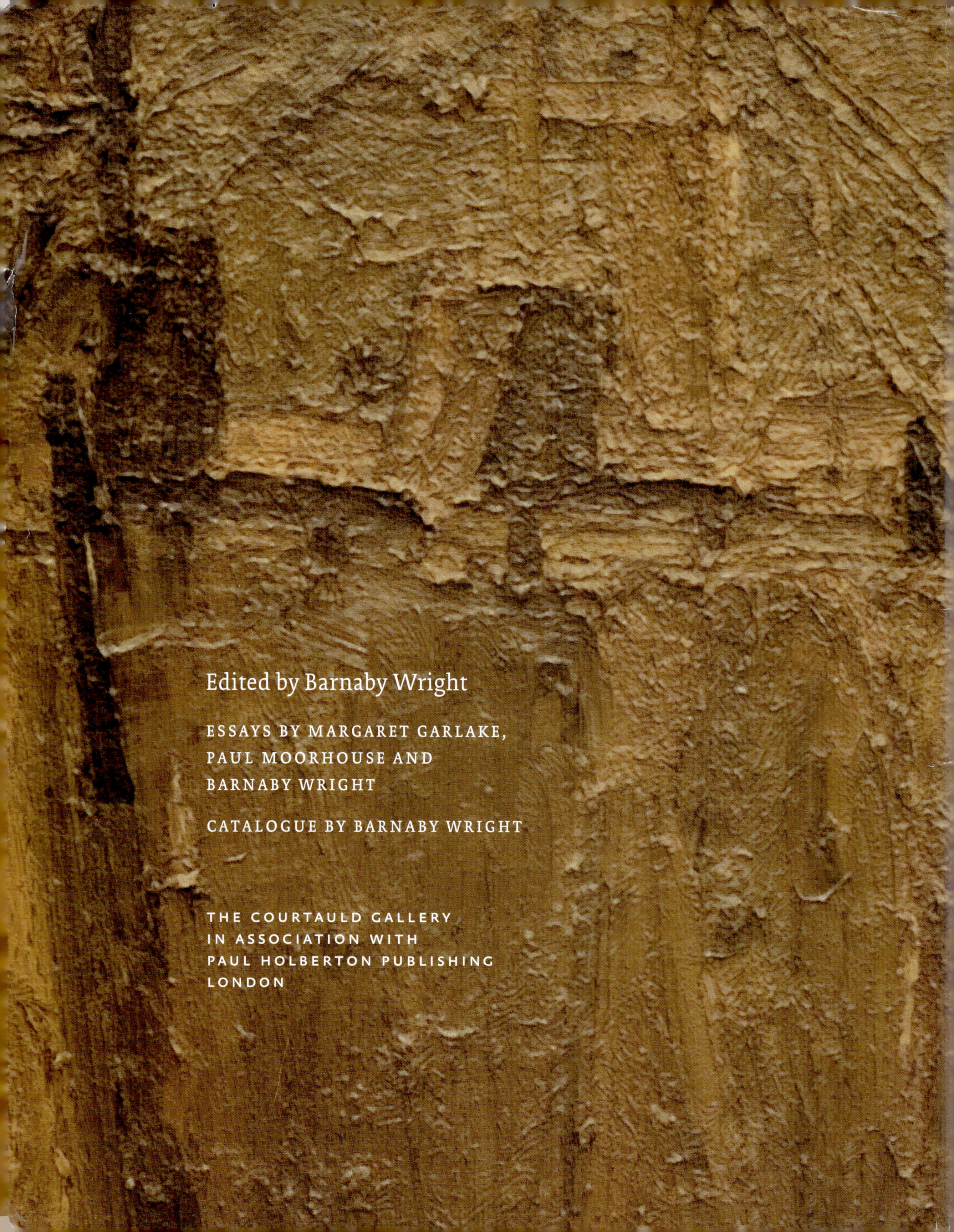

Edited by Barnaby Wright

ESSAYS BY MARGARET GARLAKE,
PAUL MOORHOUSE AND
BARNABY WRIGHT

CATALOGUE BY BARNABY WRIGHT

THE COURTAULD GALLERY
IN ASSOCIATION WITH
PAUL HOLBERTON PUBLISHING
LONDON

FRANK AUERBACH

London Building Sites 1952-62

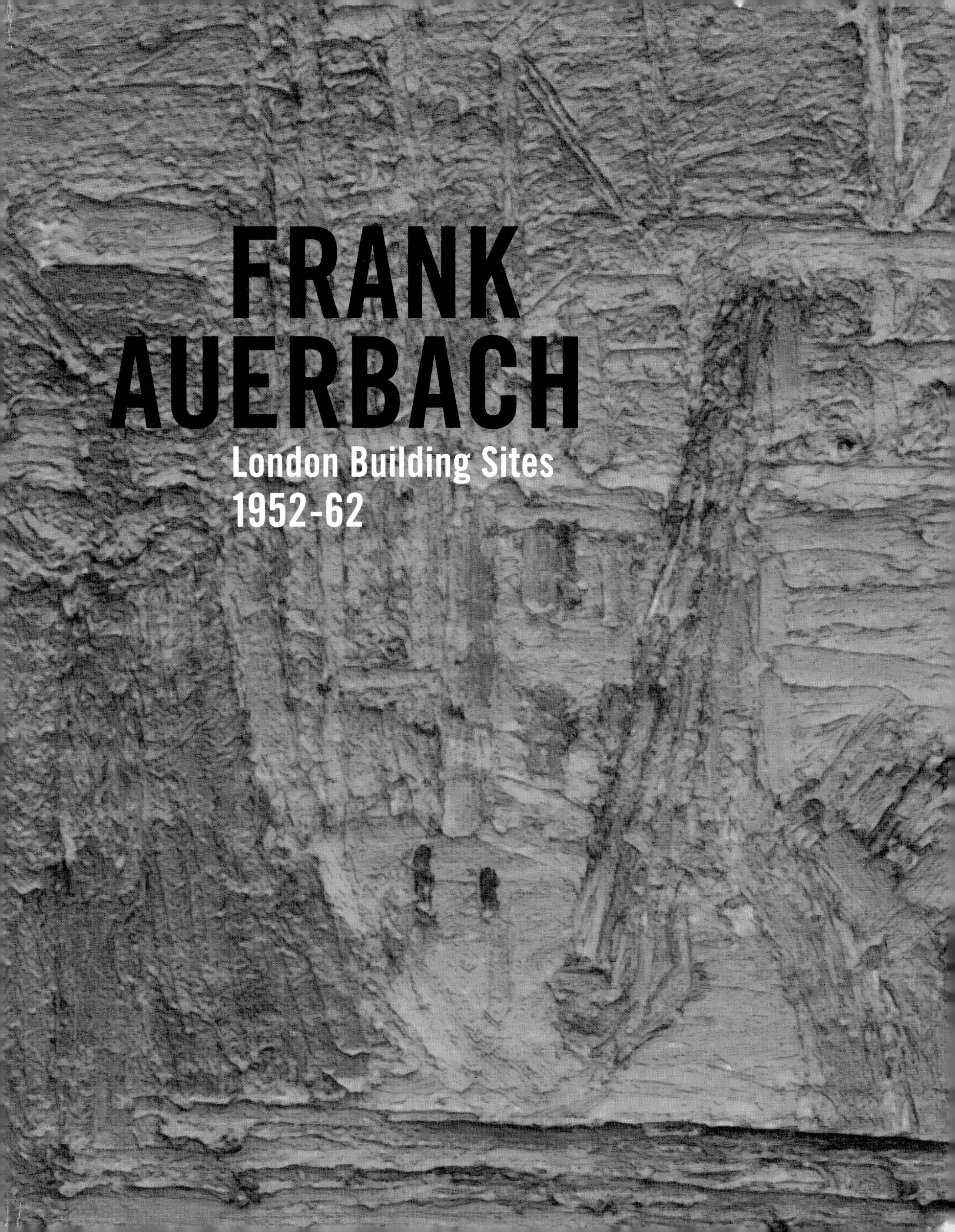

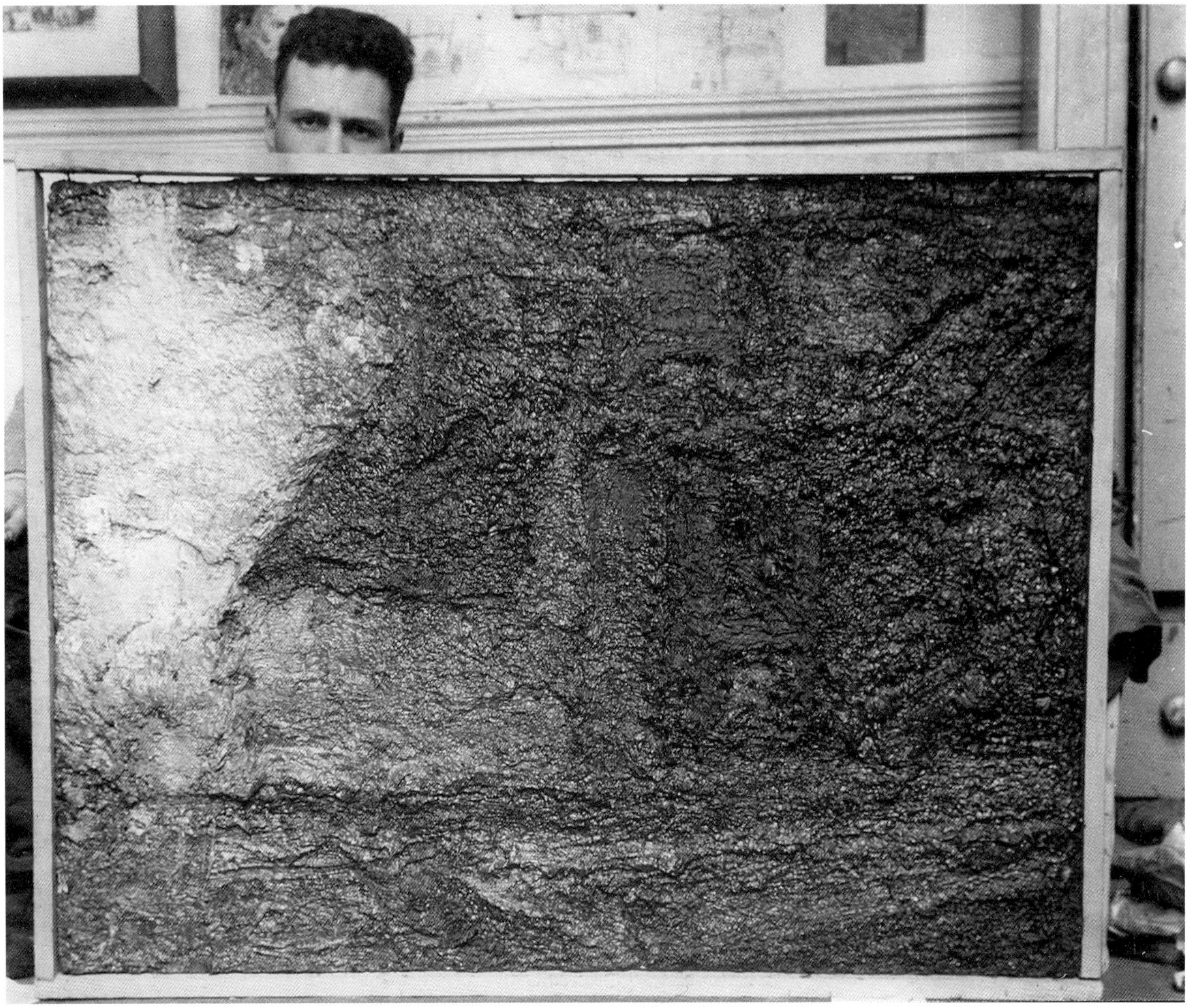

Frank Auerbach in his studio holding
Building Site, Earls Court Road (cat. 3), *c.* 1953

Beaux Arts Gallery Archive

Contents

Director's Foreword

Frank Auerbach is one of Britain's greatest living artists and we are delighted to have been able to organise this important exhibition of his London building-site paintings. It is the first exhibition to bring these paintings together as a complete group and in doing so it aims to demonstrate the importance of their contribution to post-war art in Britain. A significant amount of new research has been conducted especially for the exhibition and this catalogue shares the fruits of those endeavours. We are honoured that Frank Auerbach responded to our proposal for this exhibition with such enthusiasm from the outset and we are extremely grateful for his unwavering support for the project.

More than ever in these uncertain times, The Courtauld Institute of Art and its Gallery rely upon the generous philanthropy of those who share our vision for the future of our collections and for our unique exhibition programme. This exhibition would have been impossible to realise without the exceptional generosity of a number of supporters. Marlborough Fine Art, London, has been invaluable in bringing the project to fruition and Geoffrey Parton and his colleagues have been most sympathetic partners at every stage of the project. Once again the Friends of The Courtauld have been very generous in contributing to the exhibition and their support of this and so many other activities of both the Institute and the Gallery is greatly appreciated. The Rothschild Foundation has a long association with The Courtauld and I am very pleased that they have again given us their vital support for this exhibition. A number of supporters who wish to remain anonymous have also been crucial in making this exhibition a reality.

I would like to offer my personal thanks to David Landau, one of Frank Auerbach's regular sitters over many years and a greatly valued member of The Courtauld's Governing Board. He has given us enormous encouragement, advice and support throughout the development of this exhibition and for this, and all his work for The Courtauld, we are all immensely grateful.

Particular thanks must also go to Daniel Katz, who has given important momentum to our exhibitions and the development of our modern collections by generously funding the post of Curator of 20th Century Art at The Courtauld Gallery. This post is currently held by Barnaby Wright, the curator of this exhibition.

The Courtauld has a rich heritage of focused exhibitions, of which this latest of Frank Auerbach's work forms a distinctive part. We look forward with anticipation to many more illuminating exhibitions in the future.

DEBORAH SWALLOW
Märit Rausing Director, The Courtauld Institute of Art

Foreword

The Courtauld Gallery is immensely proud to have been able to organise *Frank Auerbach: London Building Sites 1952–62*, bringing together the artist's remarkable series of paintings of post-war London. The integrity, commitment and power of these works, which appear to be formed from the raw material of paint through sheer force of will, hold the attention and command the memory to an extraordinary degree. They are unquestionably amongst the masterpieces of post-war British painting. And yet for all the uncompromising and apparently disruptive character of the series, which pushes paint and representation to its absolute limits, these works find a deeply sympathetic home at The Courtauld Gallery.

In assembling the great collection of Impressionist and Post-Impressionist art now held at The Courtauld Gallery, Samuel Courtauld was led by the belief that the artists of what he called "the modern movement" represented not a rupture with the past but rather a profound renewal of the Western tradition of painting. This is not to say that he did not recognise the radical character of their innovations; he understood perfectly well how the Impressionists responded with urgency to the complexities of the modern world, through both technique and subject-matter. And he understood that the heroism of figures such as Cézanne, whom he admired above all others, lay in the hard-fought and often lonely battle to reinvent the language of painting with reference both to nature and the great art of the past.

Courtauld was closely involved in several initiatives to support contemporary British art, including the London Artists' Association, of which he was a co-founder, and the Contemporary Art Society, but he did not collect the art of his own time with much vigour or conviction. However, in recent years and with the essential support of a growing body of loyal and generous friends, the gallery which Courtauld so magnificently endowed has started to fulfil the promise inherent in his vision of art as an open, vital and responsive force. Gifts of individual works of art have started to add momentum to the progress of the collections, coalescing meaningfully around previously somewhat isolated bequests and single masterpieces. This path has been charted above all through a programme of exhibitions in which The Courtauld Gallery has started to explore at a level of detail sometimes overlooked by the major national collections important passages or episodes in the history of twentieth-century art.

The present exhibition exemplifies this approach. We are enormously grateful to the sponsors of the exhibition for making this project possible: several of those who have chosen to remain anonymous are long-standing supporters of Frank Auerbach's work and have embraced the ambition to bring together this powerful group of the artist's early paintings. We are grateful, above all, to Frank Auerbach for entrusting us with this project and enabling us to show these great works at The Courtauld Gallery.

ERNST VEGELIN VAN CLAERBERGEN
Head of The Courtauld Gallery

SAMUEL COURTAULD SOCIETY

The Courtauld is grateful for the significant annual support that it receives from the members of the Samuel Courtauld Society.

DIRECTOR'S CIRCLE

Anonymous
Andrew and Maya Adcock
Farah and Hassan Alaghband
Apax Partners
Christie's
Nicholas and Jane Ferguson
Sir Nicholas and Lady Goodison
James Hughes-Hallett
Daniella Luxembourg, London
Madeleine and Timothy Plaut
Paul and Jill Ruddock
The Worshipful Company of Mercers

PATRONS' CIRCLE

Anonymous
Mrs Kate Agius
Agnew's
Lord and Lady Aldington
Jean-Luc Baroni Ltd
Philip and Cassie Bassett
Charles and Rosamond Brown
The Lord Browne of Madingley
Mr Damon Buffini
David and Jane Butter
Julian and Jenny Cazalet
Mr Colin Clark
Mark and Cathy Corbett
Mr Eric and Mrs Michèle Coutts
Mark Cunningham and Her Royal Highness Princess Charlotte of Luxembourg
Samantha Darell
David Dutton
Mr Gary Eaborn and Miss Sylvia Wong
Mr and Mrs Thomas J. Edelman
Cindy Elden
Mr Roger J H Emery
Ms Veronica Eng
Mr Sam Fogg
Christine Garrett
Lucía V. Halpern and John Davies
Elaine Hansen
David Haysey
Hazlitt, Gooden & Fox
Mr and Mrs Schuyler Henderson
Philip Hudson
Nicholas Jones
Daniel Katz Ltd
James Kelly
Klein Solicitors Limited
Norman A. Kurland and Deborah A. David
Helen Lee and David Warren
Gerard and Anne Lloyd
Stuart Lochhead and Sophie Richard
Janet Martin
Mr Jay Massey
Clare Maurice
The Honourable Christopher McLaren
Norma and Selwyn Midgen
Cornelia S. Edelman Moss & James Moss
John and Jenny Murray
Mr Morton Neal CBE
Flavia Ormond Fine Arts Ltd
Mrs Carmen Oguz
Mr Michael Palin
Catherine Petitgas
Lord and Lady Phillimore
Mr Richard Philp
Mr and Mrs Herschel Post
Leslie Powell
Pam Scholes
Richard and Susan Shoylekov
William Slee and Dr Heidi Bürklin-Slee
Hugh and Catherine Stevenson
Sir Angus Stirling
Claire Tallis
Johnny Van Haeften
Mrs Elke von Brentano
Erik and Kimie Vynckier
The Rt. Hon. Nicholas and Lavinia Wallop
Jacqui and John White
Susan J. Wilen

ASSOCIATES

Anonymous
Lord Jeffrey Archer
Rupert and Alexandra Asquith
The Hon Nicholas Assheton CVO, FSA
Mrs James Beery
Bonhams 1793 Limited
Dame Diana Brittan
Henry and Maria Cobbe
Mr Oliver Colman
Emma Davidson
Hester Diamond
Simon C. Dickinson Ltd.
Mr Andrew P Duffy
David and Diane Frank
Margaret L. Ford
Mrs Judy Freshwater
The Rev. Robin Griffith-Jones
Mrs Kathryn Gyngell
Derek Johns Ltd
Jill Kastner
Timothy and Megan Kirley
Mark and Liza Loveday
Dr Chris Mallinson
Philip Mould Ltd
Roger Orf and Lisa Heffernan
Mary Oury
Jilly Parsons
The Lady Ridley of Liddesdale
Charles Rose
Edwina Sassoon
Anna Somers Cocks
Paul Stimson
Mr Robert Stoppenbach
Rex De Lisle Stanbridge
Professor Deborah Swallow
Yvonne Tan Bunzl
The Ulrich Family
The Weiss Gallery
George and Patricia White
Katherine Woodward Mellon
Nicholas and Venetia Wrigley

Exhibition Supporters

Marlborough Fine Art (London) Ltd
The Friends of The Courtauld
The Rothschild Foundation

and others who wish to remain anonymous

Acknowledgements

The Courtauld Gallery would like to thank the staff and trustees of the following institutions which have generously lent works to this exhibition: Ferens Art Gallery, Hull; Gray Art Gallery and Museum, Hartlepool; Leeds City Art Gallery; Museo Thyssen-Bornemisza, Madrid; National Gallery of Victoria, Melbourne, Australia; New Walk Museum and Art Gallery, Leicester; Pallant House Gallery, Chichester; and Tate, London. We are also extremely grateful to the many private lenders who have supported us with generous loans to the exhibition.

Catherine Lampert and William Feaver, both long-time sitters for Frank Auerbach and leading scholars of his work, have been very generous in sharing their expertise and in giving advice and support throughout the project. Geoffrey Parton and his colleagues at Marlborough Fine Art, London, have demonstrated an unwavering commitment to the exhibition. We would like to thank especially Kate Austin at Marlborough for her invaluable assistance with all aspects of the planning and research for the exhibition.

We are very grateful to Henry Lessore for assisting us so generously with the Beaux Arts Gallery Archive in his care. A number of other archives have been indispensible in research for the exhibition and we would like to extend our sincere thanks to staff of the V&A Archive and the Tate Archive and to Howard Doble of the London Metropolitan Archive. Particular thanks are due to Nick Molnar of Sir Robert McAlpine Ltd for allowing us to reproduce historic photographs from their collections. Judy Faraday and Linda Moroney of the John Lewis Archive have been extremely generous in providing historic photographs and other material from their holdings.

We are indebted to Margaret Garlake and Paul Moorhouse for their contributions to this catalogue and for their support throughout. Viktorya Vilk, a Marshall Scholar at The Courtauld Institute, has been an outstanding research assistant for the exhibition and her work has greatly enriched this catalogue.

Many colleagues at The Courtauld have contributed to bringing this exhibition to fruition and we would like to thank in particular Graeme Barraclough, Julia Blanks, Stephanie Buck, Caroline Campbell, Mary Ellen Cetra, Ron Cobb, Georgina Cox, Alexandra Gerstein, Kerstin Glasow, Katharine Lockett, Chloe Le Tissier, Joanna Selborne and Joff Whitten.

We would like to acknowledge the crucial support of the Government Indemnity Scheme, without which this exhibition would not have been possible.

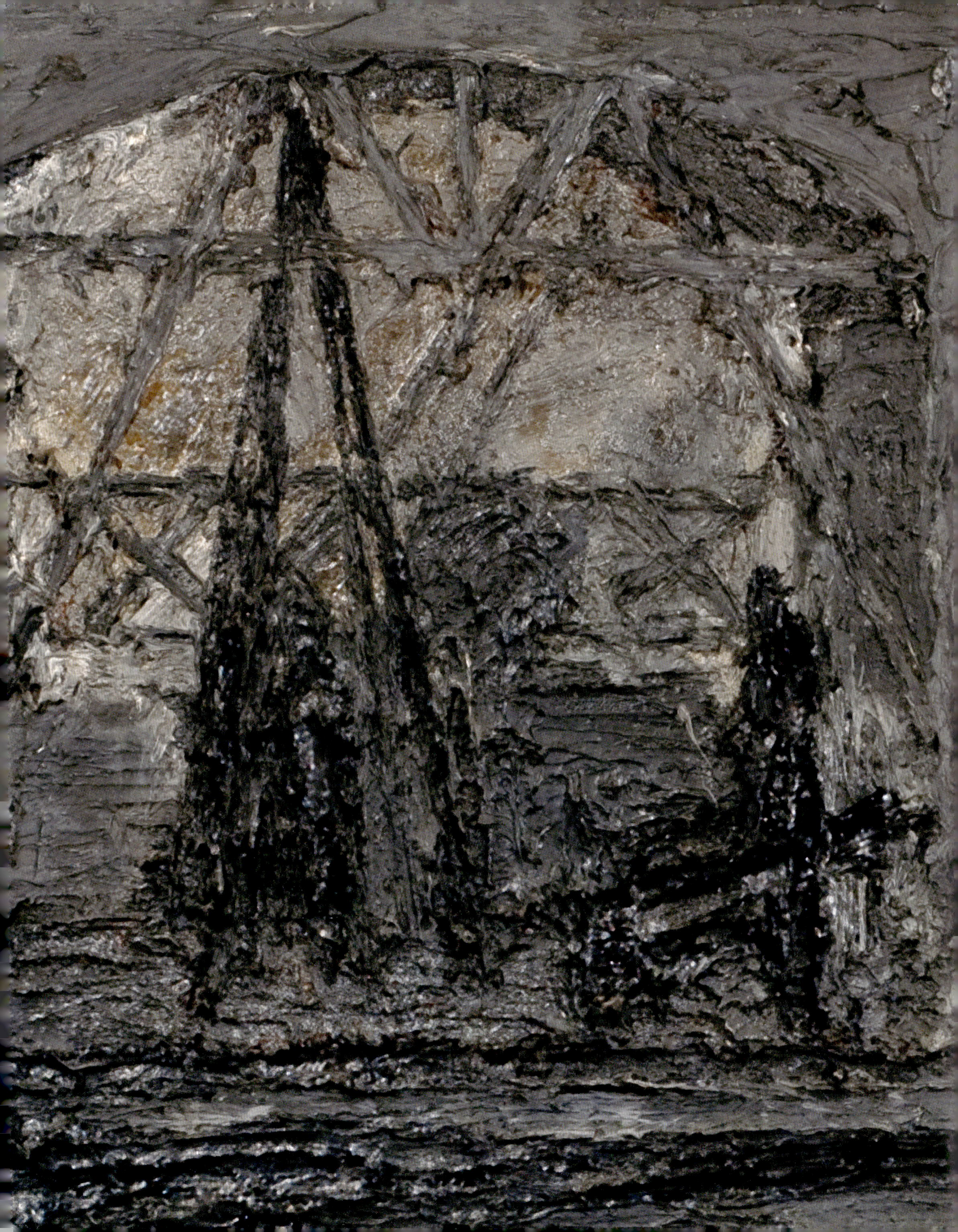

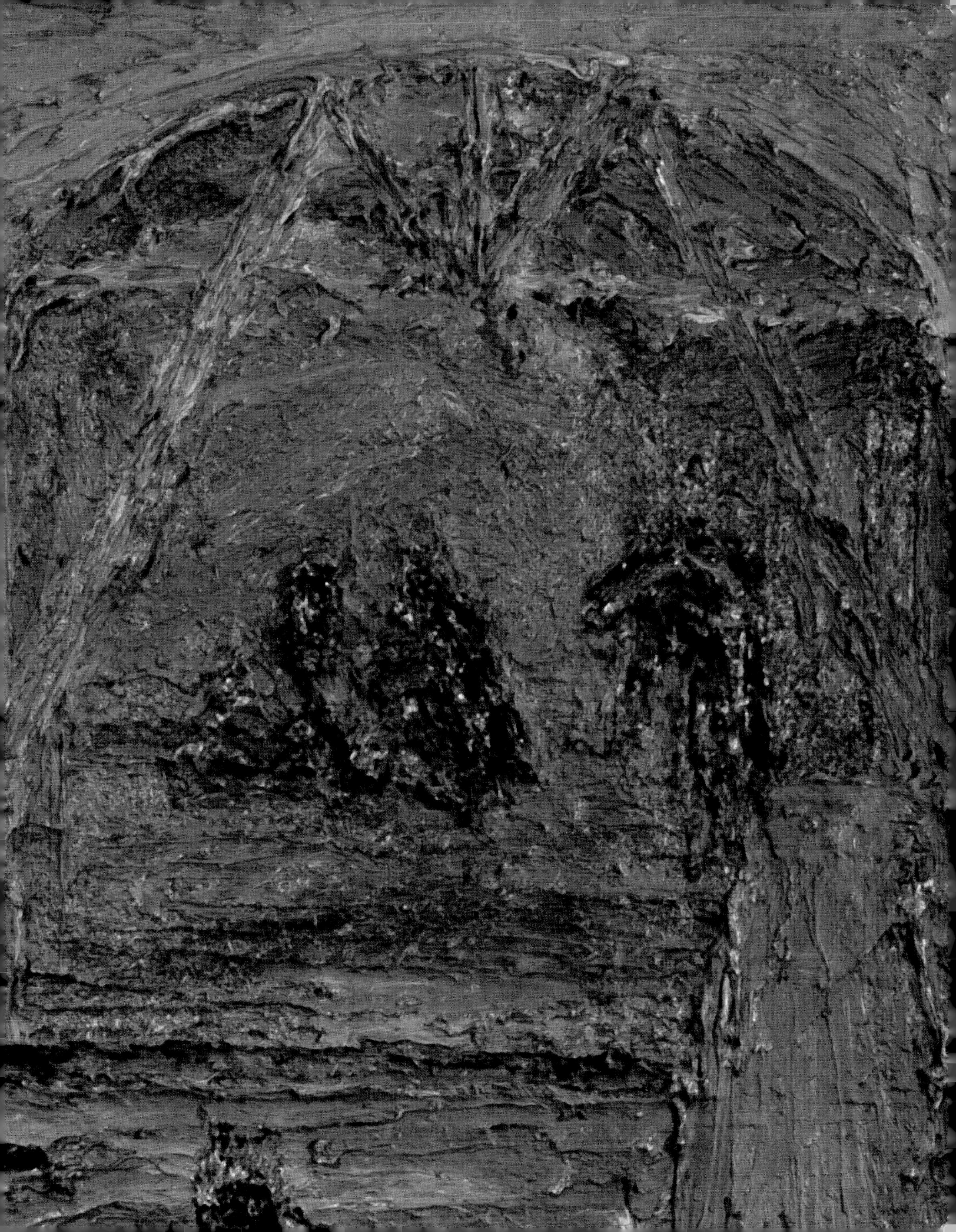

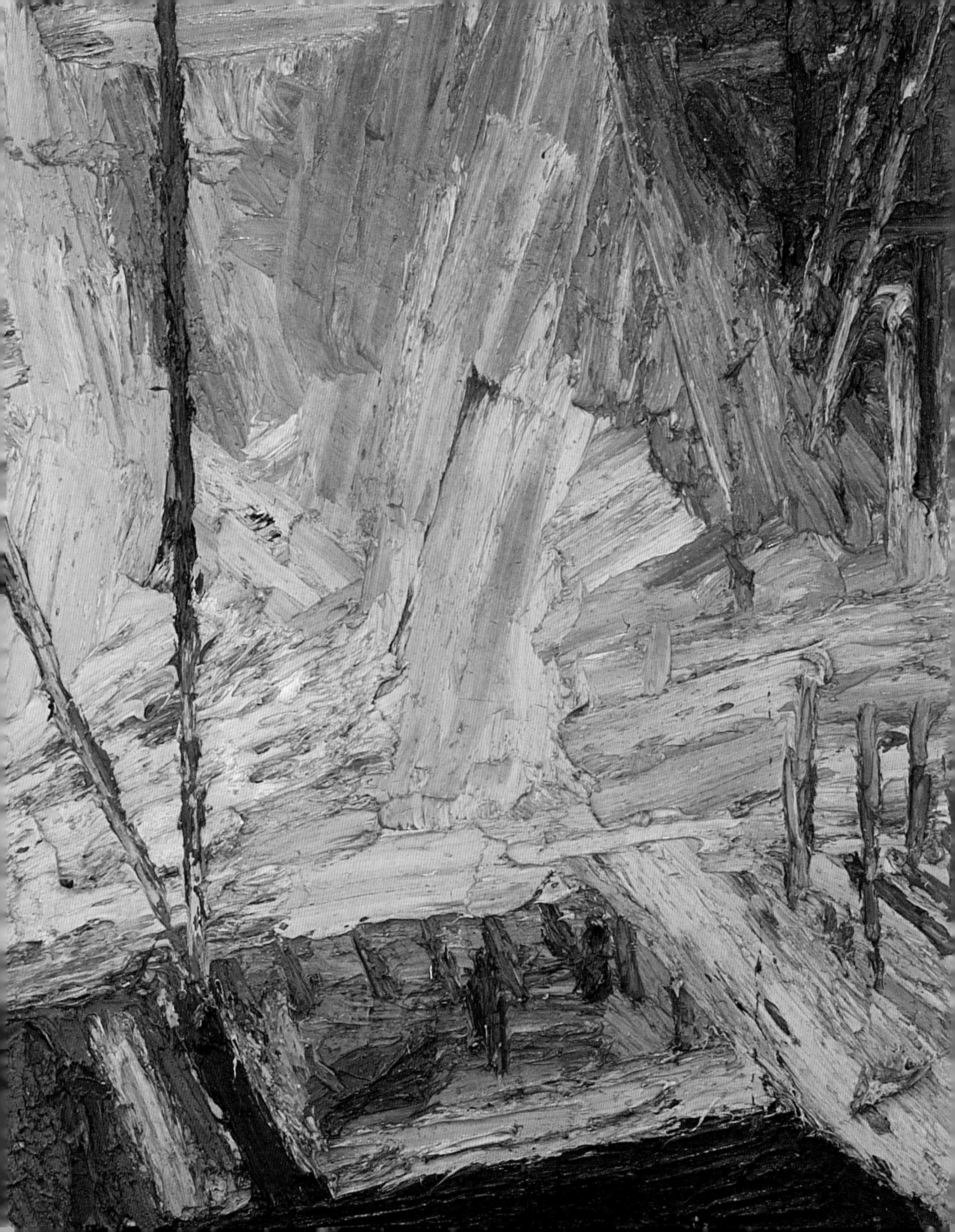

Creative Destruction: Frank Auerbach and the Rebuilding of London

BARNABY WRIGHT

"The only possible progress is to destroy ... then one's left with nothing one began with but a new fact."
FRANK AUERBACH, 'Fragments from a Conversation', 1959[1]

"Building and rebuilding cities has always been a pretty brutal business, since new construction is only possible on the ruins of something else."
LIONEL ESHER, *A Broken Wave: The Rebuilding of England, 1940–1980*, 1981[2]

Frank Auerbach's paintings of London building sites, produced between 1952 and 1962, simultaneously chart his birth and early development as an artist and the city's rebirth from the debris of the Second World War. These two births are closely connected, and this essay attempts to trace the relationships between them in order to offer an introduction to Auerbach's group of fourteen building-site paintings, which are brought together in this exhibition and publication for the first time. The project to unite the works as a complete series is based upon the conviction that a detailed exploration of their historical and artistic significance is long overdue and that their importance as a profound artistic response to London's post-war landscape demands renewed attention. An encounter with any one of them makes clear that the nature of that response is far from straightforward. Auerbach's extraordinarily thick and vigorously worked, obliterated and reworked surfaces, with paint churned, piled and deeply scored, resist easy consumption as depictions of their subject-matter. His destructive process of creation operates at a level where the conventions of representational meaning break down but are never abandoned completely.[3] The incomprehensible and unprecedented experience of seeing London ruined by war, and further dismembered across its building sites prior to reconstruction, made a profound impact upon his artistic development. The paintings are a search for a new language of painting to bear witness to these times. To understand the resonance of the works, this essay considers their relation to wider cultures of post-war reconstruction in the arts and architecture during the period, within which Auerbach's building-site paintings manifest a distinctive and indeed disruptive presence.

Auerbach's memories of his early life in Berlin, where he was born to Jewish parents on 29 April 1931, are vague and quickly culminate in his evacuation to England in 1939, shortly before his eighth birthday, as the Nazis' persecution of German Jews escalated into genocide. He spent the war as an evacuee schoolboy at Bunce Court boarding school, first in Kent and then on the Welsh Borders, with spells at cousins' and friends' houses in London during the holidays. At some point during the War he learned that his parents, Max and Charlotte Auerbach, had both been killed in a concentration camp. His temporary wartime stay in England would turn into a permanent new life as an artist in London, where he settled in

OPPOSITE
Detail cat. 9

Fig. 1
Frank Auerbach in his studio, 1963
Photograph by Jorge Lewinski

1947. In 1954, having studied at a number of different art schools, he took over a studio in Camden Town from his friend Leon Kossoff, and has worked there ever since (fig. 1). He rarely leaves the city.[4]

From the beginning of Auerbach's life in London the building sites which were multiplying across the bomb-scarred city were a constant source of fascination, as they were for Kossoff, his closest artistic companion during this period. Auerbach soon began making regular sketching trips in search of new sites across London, interspersed with time spent drawing from the Old Masters in the National Gallery (fig. 2). In a recent interview Auerbach recalled the spectacle of this extraordinary new landscape:

"Well – it's not a unique memory – but it was pitted with bomb sites and of course the bomb sites gradually turned into building sites because people were rebuilding what had been destroyed. And there was ... a sense of survivors scurrying among a ruined city .... A city fully functional is to me a somewhat formally boring collection of cubic rectilinear shapes, but London after the War was a marvellous landscape with precipice and mountain and crags, full of drama formally, and I think it was that ... I mean if I'd lived in the mountain landscape, which I'm very pleased to say I didn't because that had been mined in the previous century, I might have been drawn to try and paint that. But

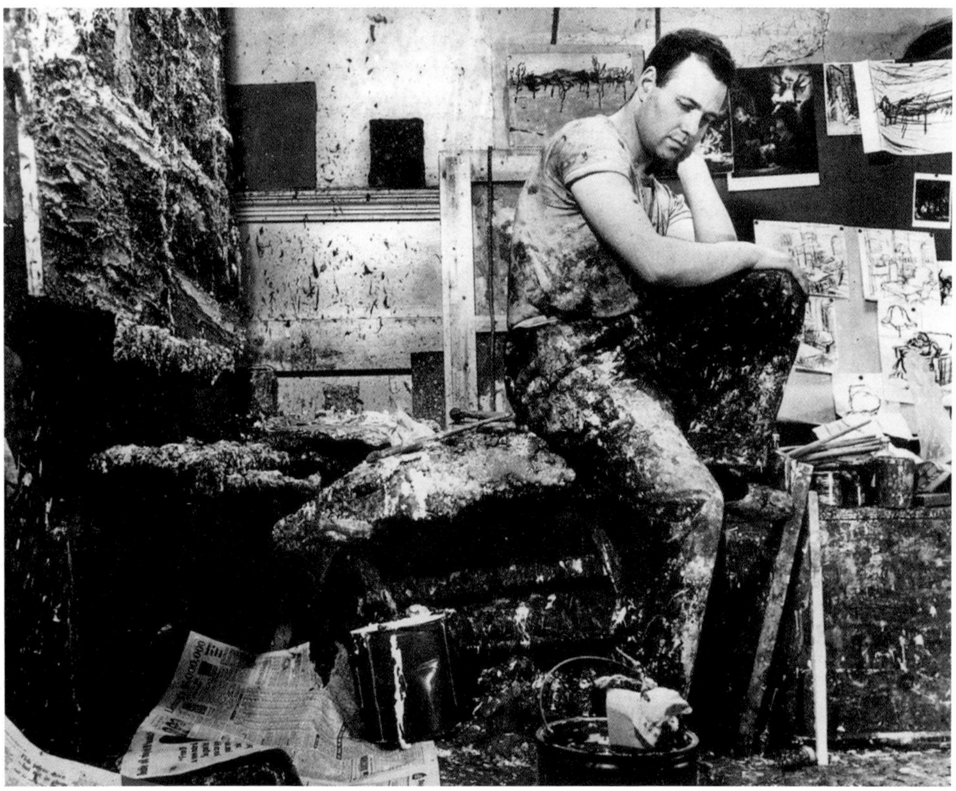

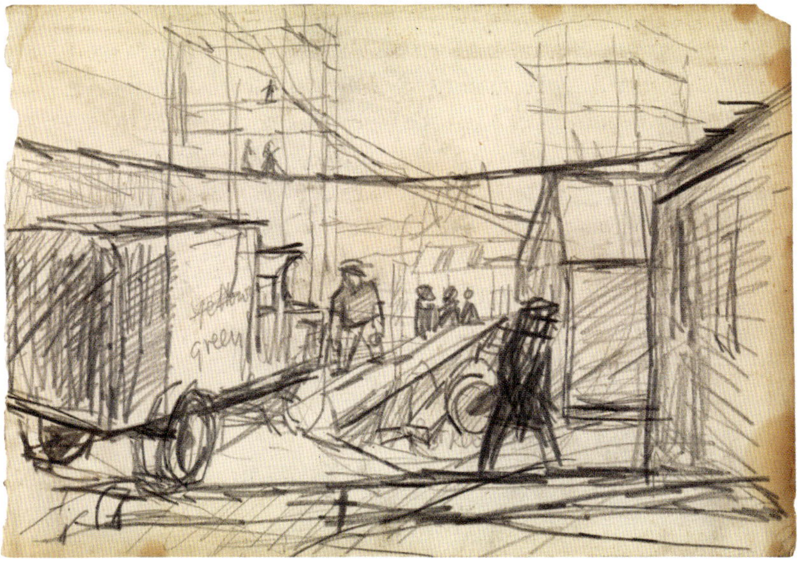

Fig. 2
Frank Auerbach, *Sketch of a Building Site*
c. 1952
Pencil on paper, 17.7 × 25.8 cm
Private collection

living in London it seemed mad to waste the opportunity not to take notice of the fact that there were these marvellous images, compost for images all around one .... Being in Paris, you'd be haunted a little bit if you started trying to paint Notre Dame or a view of the Seine by the fact that other people had painted it before, but the London I knew and saw hadn't really been painted ... it was a new phenomenon."[5]

For Auerbach, the excitement of this unknown phenomenon of London turned into a wild mountain landscape was mixed with anxiety and fear of the sites themselves: "I remember in those days going to places where I was scared to stand and drawing .... On these planks people wheeled wheelbarrows totally confidently, I would sit down and edge my way along them in order to do my drawings. I've got no head for heights at all."[6] What drove him to confront his fears was the challenge of finding a new artistic language with which adequately to express this unprecedented subject.

Bombed London and Neo-Romanticism

The question of how to comprehend and picture the unexpected and unregulated cityscape, newly created by the destructive power of the German bombs, had found responses even before the air raids had abated and well before Auerbach settled in London. J.M. Richards's *The Bombed Buildings of Britain* was published in 1942 and through numerous evocative photographs gave an account of the damage wrought to buildings in London and elsewhere between 1940 and

1941 (fig. 3). In his foreword Richards makes the case for the devastation to be understood as part of a rich history of the aesthetic power of ruins from: "Storm and lighting, the death-watch beetle, Cromwell's troopers ... and now German Bombs." He continued:

"At first mention it may appear unfeeling that the connoisseur of ruins should regard as material for objective appraisal scenes which for most of us symbolise the horror of lives lost and irreplaceable treasures destroyed. But it is surely the very intensity of the symbolism they are charged with that justifies the claim of ruined buildings to be looked at for the sake of what they are at the moment ... for the sake of the intensely evocative atmosphere they possess in common with all ruins, which gives them an architectural vitality of their own; and frankly for their beauty."[7]

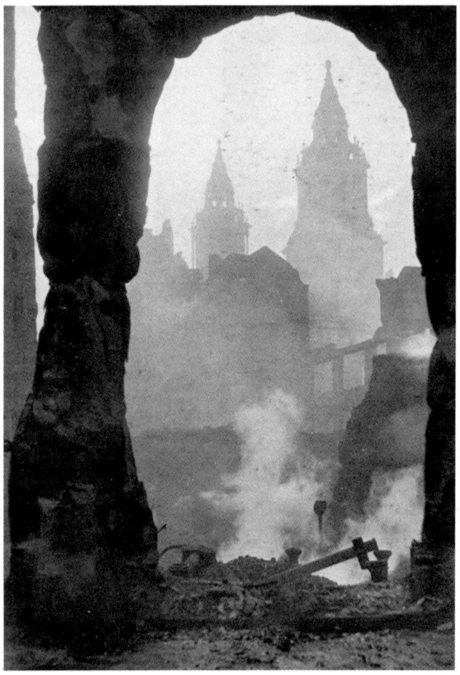

Fig. 3
Photograph of ruined buildings around St Paul's Cathedral, from J.M. Richards, *The Bombed Buildings of England*, 1942

Richards's account of "the romantic appeal" of "the architecture of destruction" was part of an emerging and broadly coherent response to the aftermath of the bombing by influential members of the architectural and artistic establishment, who wrested a neo-Picturesque and neo-Romantic aesthetic from the devastation of the Blitz. Richards gives almost a blueprint for how this could be visualised in a given ruin. He undoubtedly had in mind the work of John Piper and Graham Sutherland (see fig. 4).[8] Their neo-Romantic paintings of bombed buildings in London and elsewhere from 1940 onwards were sponsored by the Ministry of Information's War Artists' Advisory Committee (WAAC), under the control of Kenneth Clark, then director of the National Gallery. Alongside Henry Moore's shelter drawings, these works became a prominent visualisation of the effects of war in Britain and were shown at hugely popular exhibitions at the National Gallery and elsewhere throughout the period of hostilities. Their fusion of avant-garde languages of art, notably drawn from Cubism and Surrealism, with a revival of early nineteenth-century British Romantic traditions of painting (among a range of other influences) "came to embody", as Virginia Button concludes, "the ideals of British liberal culture, now threatened with extinction by totalitarian forces".[9]

The neo-Romantic work given public prominence by the WAAC's exhibitions offered a response to the destruction of war in which the bombs could be seen to have created a new beauty within which lay the seeds of reconstruction itself. In his 1948 book, *Buildings and Prospects*, Piper argued for the beauty of the bombed ruins, with their picturesque qualities of irregular contours and contrasts of scale and form, to be the inspiration for future rebuilding efforts. He further argued that the work of pre-war avant-garde painters could provide aesthetic direction:

"The incorporation of Picasso and Matisse, Ernst and Miró into our visual philosophy may mean an uncomfortable stretching of the word 'picturesque' to embrace our beliefs .... They prophesied the beauty as well as the horror of bomb damage, and as visual planners they are at the moment unrivalled. Bomb damage has revealed new beauties in

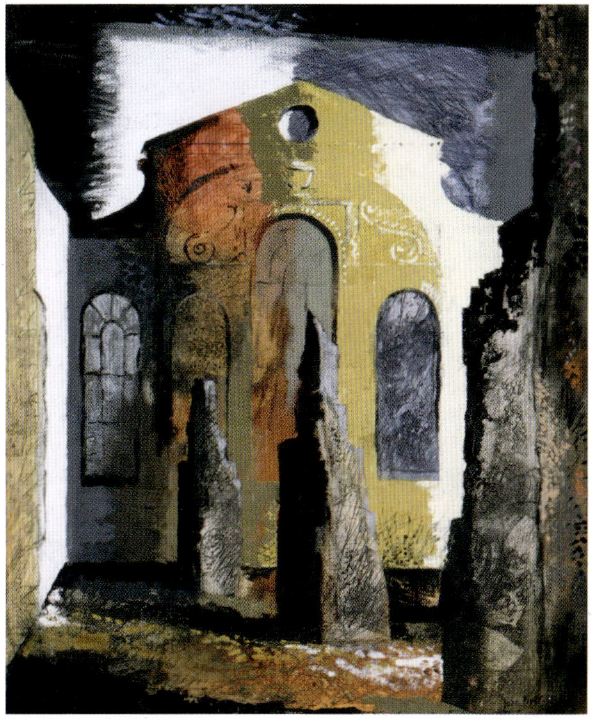

Fig. 4
John Piper, *Christ Church, Newgate Street, London*
1941
Oil on canvas, 75.3 × 63.1 cm
Museum of London

unexpected appositions as a rich source of information for the planner who would retain picturesque elements from the past."[10]

Piper's argument fed into the vigorous architectural debates surrounding reconstruction. As Margaret Garlake demonstrates in her essay in this catalogue, notions of a modern picturesque played an important role in shaping the rebuilding of London in the post-war period, in competition with attempts to assert a hard-edged Modernist architectural vision, and with those advocating an historicised, neo-Georgian aesthetic. But although opposing voices constituted the architectural discourse of reconstruction in Britain, the neo-Romantic and neo-Picturesque vision remained the dominant artistic response to London's devastated landscape at the time Auerbach arrived in the city.

Auerbach's Vision of London

It is telling that from the outset of his sketching trips around London Auerbach was not attracted to the bombsites or the ruined buildings themselves but to construction sites. This was territory that had not been resolved into a neo-Romantic idiom and did not lend itself readily to past idealisations as an aesthetic of ruins. Indeed, Auerbach's favoured stage of a site was when the building had only just begun to emerge from the ground and the contrast between the chaos and formlessness of the excavated earth and the early suggestions of architectural order was at its greatest point of tension. This separates his work from the glorification of the spectacle of steel-framed constructions in photographic accounts of the birth of modern cities in the later nineteenth and early twentieth century, from the Eiffel Tower to the skyscrapers of New York, and was still the focus of many photographs of London's post-war rebuilding (see, for example, fig. 36). In 1950 this modernist sensibility would find one of its most famous painted expressions in Fernard Léger's *The Constructors* (fig. 5). But in the main building sites were not subjects that had been exploited by major modern European artists of the period and for Auerbach this was one of their attractions.

Although Auerbach had been making sketches of building sites from perhaps as early as 1950, whilst a student at St Martin's College of Art, he had either not tried to turn these into paintings or else the attempt had failed. Though he destroyed most of these drawings, a few survive, including a squared-up study of workmen constructing a roof which he started as a painting but subsequently abandoned (fig. 6). His breakthrough came in September 1952, while working on *Summer Building Site* (cat. 1), which he had been struggling with for months. He has recounted the circumstances several times subsequently, but first in his 1959 text, 'Fragments from a Conversation', published in the first issue of the short-lived literary journal *X*: "I could tell you the date at which I reached it, finding the image – strange and new to me. In the morning I'd been working, very, very conscientiously, painting a building site . . . suddenly I was conscious of something underneath it . . . this building site, I'd done it again and again, I knew it intimately . . . and then there was suddenly the image underneath it . . . . I'd destroyed all the reminders (that is, of painting) to get a unique thing . . . . It began to operate by its own laws."[11] Some fifty years later he still cites that day as being "the beginning of my life as a painter".[12]

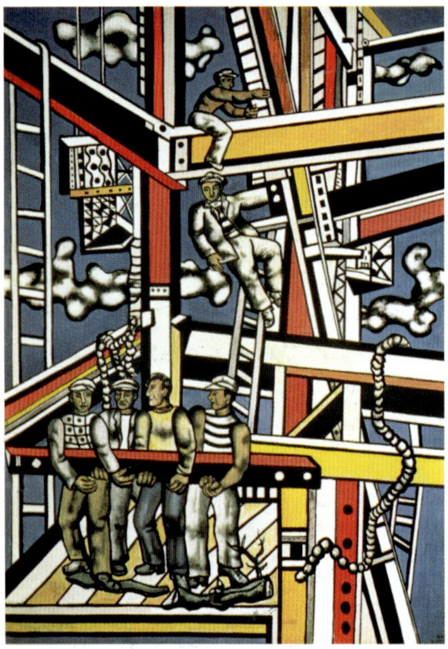

Fig. 5
Fernand Léger, *The Constructors*
1950
Oil on canvas, 300 × 228 cm
Musée Fernand Léger, Biot

The breakthrough for Auerbach of *Summer Building Site* was the experience of pushing a painting beyond the limits of artistic convention and straightforward representation by painting and repainting to achieve a new and more vital expression. As Paul Moorhouse discusses in his essay for this catalogue, David Bomberg's life-drawing classes at the Borough Polytechnic, which Auerbach attended with Kossoff, were influential in the development of Auerbach's persistent way of working. Bomberg encouraged him to explore his subject with dogged determination. "What happened was that people would draw," he recounted of Bomberg's classes; "as soon as they seemed to be drawing in a way that was bitty, affected, mannered or using a cliché learned from art, he would refer them back to the model and . . . suggest a total destruction, and they would destroy it and go on; and probably where they were least aware of having done a picture he might stop suddenly and say, 'There is some quality in this form'."[13]

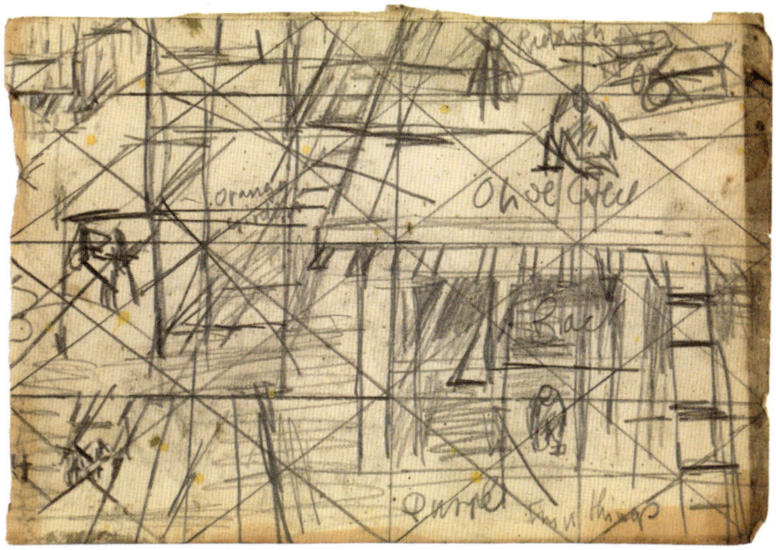

Fig. 6
Frank Auerbach, *Sketch of a Building Site with Workmen on a Roof*
c. 1950–52
Pencil on paper, 17.7 x 25.8 cm
Private collection

Auerbach would soon take this destructive process of creation well beyond Bomberg's teaching, driven by a desire to create works which came close to achieving the aesthetic power of the Old Masters he most admired in the National Gallery. It soon became typical for Auerbach to work on a painting intensively for months or even years, repeatedly working over the previous day's efforts, obliterating passages and rephrasing them until the sum of these destructions began to add up to a coherent new image. *Building Site, Earls Court Road* (cat. 3) epitomises how in the process the paint surface of a work gradually accrued until it stood out in relief, often an inch or more thick. By the time he was satisfied that a work was finished it had amassed an archaeology of paint buried in successive layers beneath the peaks and valleys of its heavily worked surface. Each painting was rooted in close observation of the sites themselves, recorded during sketching trips to a given site. He would visit a site perhaps five or six times over a relatively short period, usually at an early stage in the construction, capturing what he needed before the works moved on and the scene had changed completely. Throughout the process of painting selections of the numerous sketches remained pinned up on his studio wall and were a constant reference point. In some works, such as *Building Site, Bruton Street* (cat. 2) and *Maples Demolition* (cat. 12), a complete view sketched in one drawing became the full composition of the painting. In others, such as *Oxford Street Building Site I* and *II* (cat. 11a and 11b), different elements from a number of sketches were worked into the paintings. Despite this, representational details could be smothered by successive layers of paint or obliterated by the deep gouging of a wide brush dragged through the encrusted surfaces. Legibility was subject to sacrifice if it brought Auerbach closer to achieving what he describes as "formal grandeur" or "a secret

Fig. 7
Frank Auerbach, *Railway Arches, Bethnal Green*
1958
Oil on canvas, 111.7 × 139.7 cm
Private collection

internal geometry" which gradually emerged after months of labour and struggle.[14]

London's construction sites remained a constant theme in his work for the next decade, alongside his other major preoccupations, the human figure and the park landscape of Primrose Hill (see fig. 26). He also made a small number of paintings of other urban motifs which interested him, such as Smithfield Market and the railway arches in Bethnal Green (fig. 7). He crossed the city in search of ever more exciting building-sites, ranging from relatively small-scale residential constructions on the Earl's Court Road and Portobello Road in the early years (cat. 1, 3 and 5) to enormous sites such as the Shell Building development on the South Bank at the end of the decade (cat. 7–9). Seen as a complete group of fourteen paintings, the series amounts to a mapping of a number of the most symbolically resonant locales, such as the area around St Paul's Cathedral (cat. 6) and the rebuilding of John Lewis's famous department store on Oxford Street (cat. 11a and 11b). It also covers the major types of building project undertaken during the period, from social housing schemes such as the Regent's Park Estate (cat. 4) to the rise of office block developments like that on Victoria Street (cat. 10) – monuments to the power of a new breed

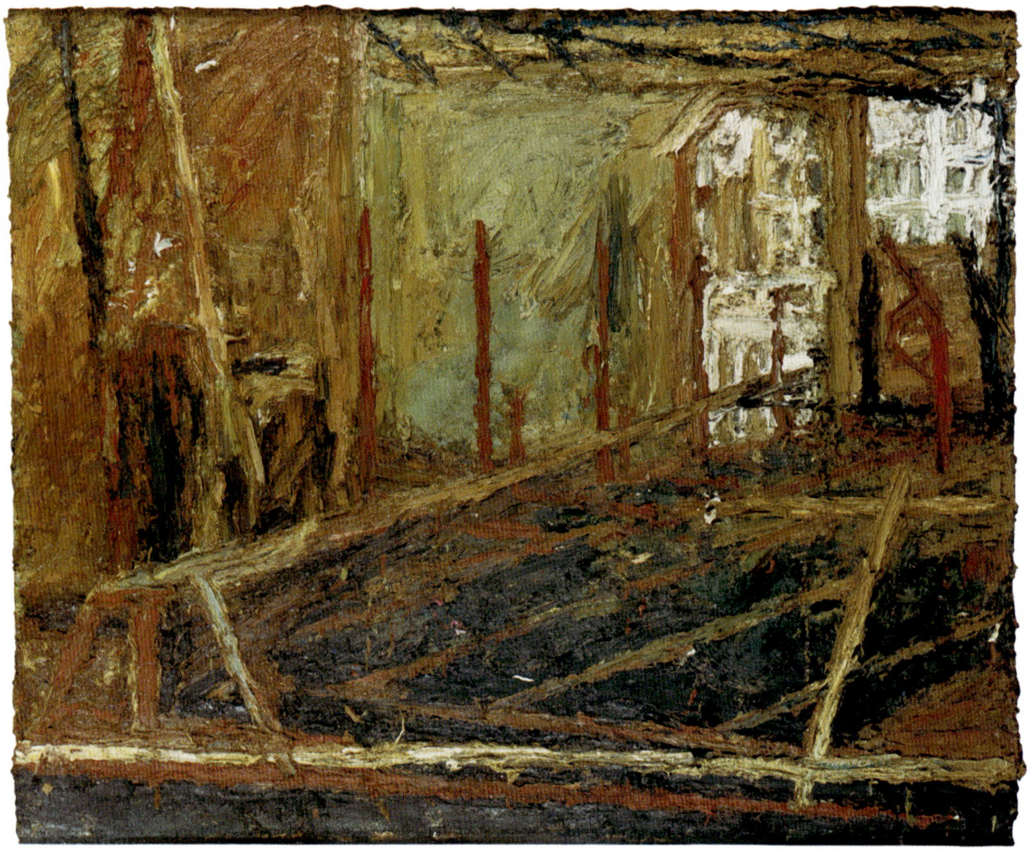

Fig. 8
Leon Kossoff, *Building Site, Victoria Street*
1961
Oil on canvas, 122.5 × 184.8 cm
Arts Council Collection

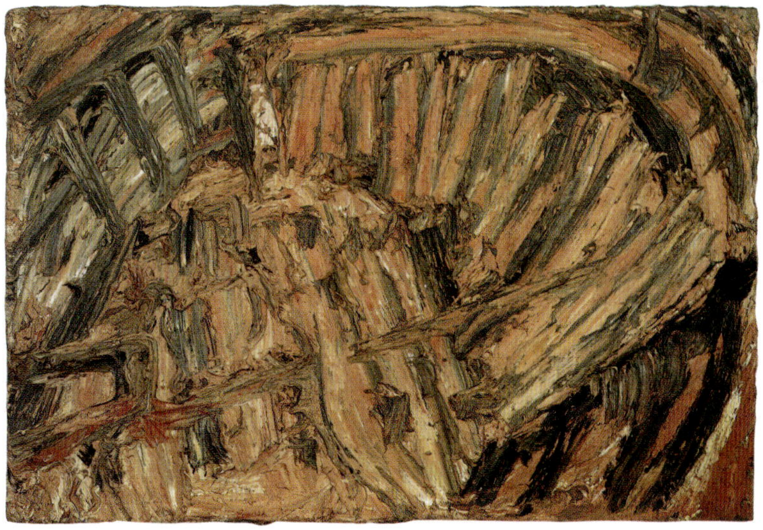

of commercial developer. It has been a concern of this exhibition to establish the specific locations and history of the sites Auerbach painted in order to place the works more fully within the historical fabric of their time (see the catalogue section). However, any impression that Auerbach intentionally selected sites for their symbolic or historical significance is fundamentally to misunderstand his motivations (see cat. 6). "Was it Théophile Gautier who said painters are basically beasts?" he asked in a recent interview about the building-site paintings. "I think I just wandered around and drew what I wanted to draw. I didn't have any general ideas about it .... Subject for me has always been a physical thing. Intimacy matters, but programmes don't – for me."[15]

That Auerbach's developing and radically unconventional approach to painting was closely aligned with London's building sites is significant in other ways. It is clear that the paintings ran against the grain of previous depictions of the war-torn city, especially the symbolic picturesque aesthetic that was the public face of neo-Romanticism. Auerbach's unrelenting earth tones and overwhelmingly dense paint-surfaces indeed seemed to reviewers of his first solo exhibition at the Beaux Arts Gallery in 1956 to be decidedly anti-picturesque. For those unsympathetic to his work this was grounds for criticism: "His technique is quite terrifying," claimed one critic. "It is by its very nature repulsively messy."[16] But for others, such as David Sylvester, who recognised the revolutionary character of his paintings, Auerbach's technique "extended the power of paint to re-make reality".[17] Auerbach had found a new language to express his response to London's cityscape. Apart from Kossoff, whose commitment to building sites as a subject of his art paralleled Auerbach's own (see fig. 8), few other modern artists in London painted them at all and none with such intense devotion.[18]

"What happened on the building site?"

This is a question posed but unanswered in Auerbach's 1959 'Fragments from a Conversation'.[19] It went unanswered, perhaps, because the experience of London's post-war rebirth witnessed by him on its building sites evaded tidy explanation or clear comprehension and was without precedent. This was what gave them their power for Auerbach. Many accounts of the aftermath of the bombing in London attest to the physically and psychologically disorienting experience of the city transformed out of all recognition. Peter Ackroyd suggests that "it seemed to many that the tangible and textural history of London was without meaning, if its glory could disappear in a night; it was too fragile and frail to be relied upon".[20] The writer Bryher, struggling to make sense of London's ruins towards the end of the war, concluded: "This was what the war was making of us, it was giving us hallucinations and driving us back to the blank spaces of our beginnings. My mind was doubling round and round on itself like a bird off course."[21] The experiences of those who had lived through the Blitz were, inevitably, different from those whom, like Auerbach, came to London shortly after the war and had only limited exposure to the bombing itself.[22] However, common to many Londoners, new and old, was the vertiginous character of life in the post-war city, where for a short period concepts of normality, convention and reality were in free fall. "It was improvisatory," as Auerbach recalled.[23] It is in this context that Auerbach's radical challenge to conventions of representation in his building-site paintings needs to be understood.

'Fragments from a Conversation' is a text of these improvisatory and unstructured times. It consists of extracts from a verbal conversation with Auerbach written down by the author Elizabeth Smart, compiled as a stream of consciousness and published under Auerbach's name. It is clear enough from these fractured statements that Auerbach's destructive process of painting was driven by a compulsion to create an image that was profoundly new: "You've got to be left with something that wasn't there before ... it's got to be a revelation".[24] This desire was, on one level, the expression of his youthful hunger to assert himself as a modern artist at a time when a "curious freedom" was in the air.[25] But it can also be seen as fundamentally entwined with the building sites themselves. As the pair of quotations which open this essay attest, Auerbach's enactment of a creative destruction in the studio paralleled that being carried out across London's building-sites as bombed ruins were bulldozed and buried and new structures built upon them. An analogy between his thick paint surfaces and muddy palette and the earthworks of the building sites has often been noted. As Nevile Wallis put it in his review of the artist's 1959 Beaux Arts exhibition, "Mr Auerbach's recent paintings here of building sites illustrate one of the various imaginative ways of seeing the new London ... he [peers] down into the building craters ... and so bulldozes his thick, glutinous pigment across the canvas to convey an exact sense of the intractable mounds of clay, and the thrust and strain of tackle".[26] But it was not Auerbach's intention to forge such connections: "I didn't want them to be thick, they just came out thick", as he has often said.[27] And his use of cheap earth colours was driven by economic necessity and not by mimetic sensibility – as it was, one critic noted wryly that the amount of paint he used "should delight the heart of an artist's colourman".[28] Looking beyond this, it is the theme of destruction and creation itself that offers the most compelling connections between the paintings and their subject-matter.

Creation and Destruction in the Modern City

The impulse to destroy in the progressive pursuit of the new was a long-standing pre-war Modernist ambition, often associated with urbanism and the city. It had been expressed to varying extremes both in the arts and in architecture, from Futurist Umberto Boccioni's revolutionary 'manifesto' painting *The City Rises*, 1912 (Museum of Modern Art, New York) to Le Corbusier's *Plan Voisin*, 1925, which proposed bulldozing most of central Paris to make way for a series of grid-pattern tower-blocks. In post-war London this ambition had in part been achieved by the bombing, allowing long-held fantasies of a *tabula rasa* suddenly to turn into a reality. Even traditionalist architects and artists engaged in reconstruction shared something of this Modernist dream, since even for those most sensitive to maintaining continuity with the past the opportunity to contribute to the rebuilding of one of the world's greatest cities was one to be seized. For all the various efforts to respect London's history in the planning of its future, the numerous grid-patterned tower-blocks, for both housing and offices, that sprang up across the city were a highly visible expression of Modernism's at least partial triumph in the rebuilding of London. The grid, as Rosalind Krauss put it in another context, was "emblematic of the modernist ambition".[29] The ideals of reconstruction, in their varied expressions, all inevitably aimed at the reparation of the trauma of hostilities through the application of new and progressive architectures which would reorder the city and sweep away the debris of war – a task for which Modernist practitioners, of varying degrees of radicalism, were well rehearsed.[30] The ambition had been powerfully expressed in Abram Games's wartime propaganda poster (fig. 9), which presents the Modernist ideal, in the form of Berthold Lubetkin's Finsbury Health Centre building, replacing a disease-ridden, bombed-out slum.

However, Auerbach's artistic impulse to destroy and create anew produced results of a very different order from these Modernist ideals. The contrast between *Building Site, Bruton Street* and its preparatory drawing (cat. 2 and 2 (i)) is among the most telling. It demonstrates how,

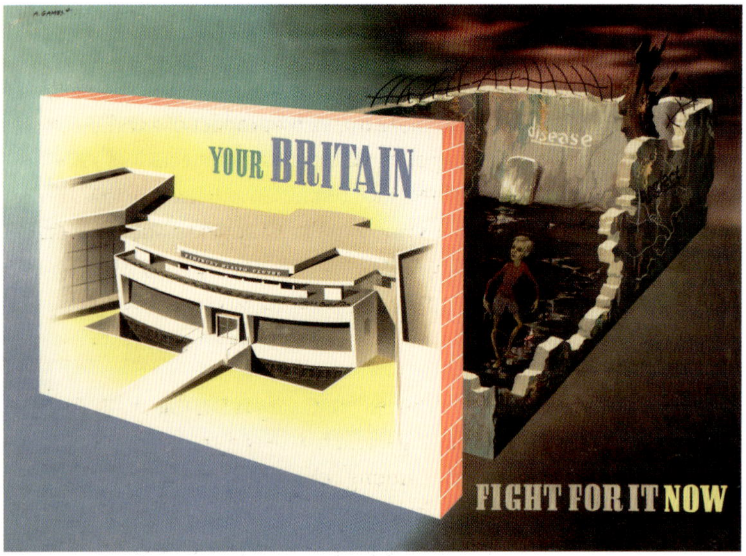

Fig. 9
Abram Games, *Your Britain – Fight For it Now*
1941
Poster
Imperial War Museum, London

in the long process of painting, the imposing Modernist structure of the Time-Life Building's steel-framed grid, recorded in the drawing, became overwhelmed by the sheer weight and thickness of Auerbach's paint. The inherent formlessness of paint as a material substance is asserted to the point where it subverts the rigid structuring geometry of the framework. In a work such as *Building Site near St Paul's* (cat. 6), in which a grid of steel girders is the principal subject of the painting, Auerbach's repeated inscribing of the lines of these girders is taken to the verge of obliteration within the mass of paint. Far from confirming architectural order and structure, the rendering of the framework again underscores the ability of paint to destabilise and subsume form. In two of the Shell Building site paintings (cat. 8 and 9), the areas of the construction site that most interested him were the excavated mountains and canyons of earth, whose broad expansiveness he played off against the linear forms of the crawler cranes and equipment: these offer a sense of only temporary structure and visual order amidst the otherwise incomprehensible muddy mass.

Form, Formless and Reparation

The best informed of Auerbach's early reviewers contrasted his work with that of Jean Dubuffet and Jean Fautrier, whose so-called 'matter paintings' were constructed from thickly applied pigment, sometimes with the addition of mud, sand, cement and other foreign materials (see fig. 10). They did so in order to set Auerbach apart from these artistic practices and from the contexts of Tachisme or L'Art Informel in post-war Paris. John

Fig. 10
Jean Dubuffet, *Le voyageur sans boussole*
1952
Oil on board, 118.5 × 155 cm
Musée national d'Art moderne – Centre Georges Pompidou, Paris

Berger quickly dismissed any connection between his technique and "Dubuffet's mud tricks",[31] and David Sylvester noted that whereas for the Frenchmen material thickness was contrived as a symbolic end in itself, for Auerbach "the thickness is no more than a by-product of another purpose".[32] However, the building site paintings were especially pertinent to art and ideas that were concerned with oppositions between 'form' and the 'formless' (or *informe*). This was broadly an antagonism between formalist Modernism (in a variety of guises) and its dissenters, such as those associated with L'Art Informel. It constituted one of the major discourses in post-war art, centred upon Paris but with much wider ramifications.[33]

The writings of the 'dissident' Surrealist, Georges Bataille, helped fuel the debates at their extreme edge, and have been connected with the work of Dubuffet and Fautrier among others, including Auerbach's friend Francis Bacon.[34] Bataille's thinking also became entwined in the wider debates of existentialism in the post-war years.[35] One of Bataille's most influential arguments was that humanity had become enslaved by the increasing dominance of architecture, which he defined as being all systems and categories of knowledge and thought: "that which orders

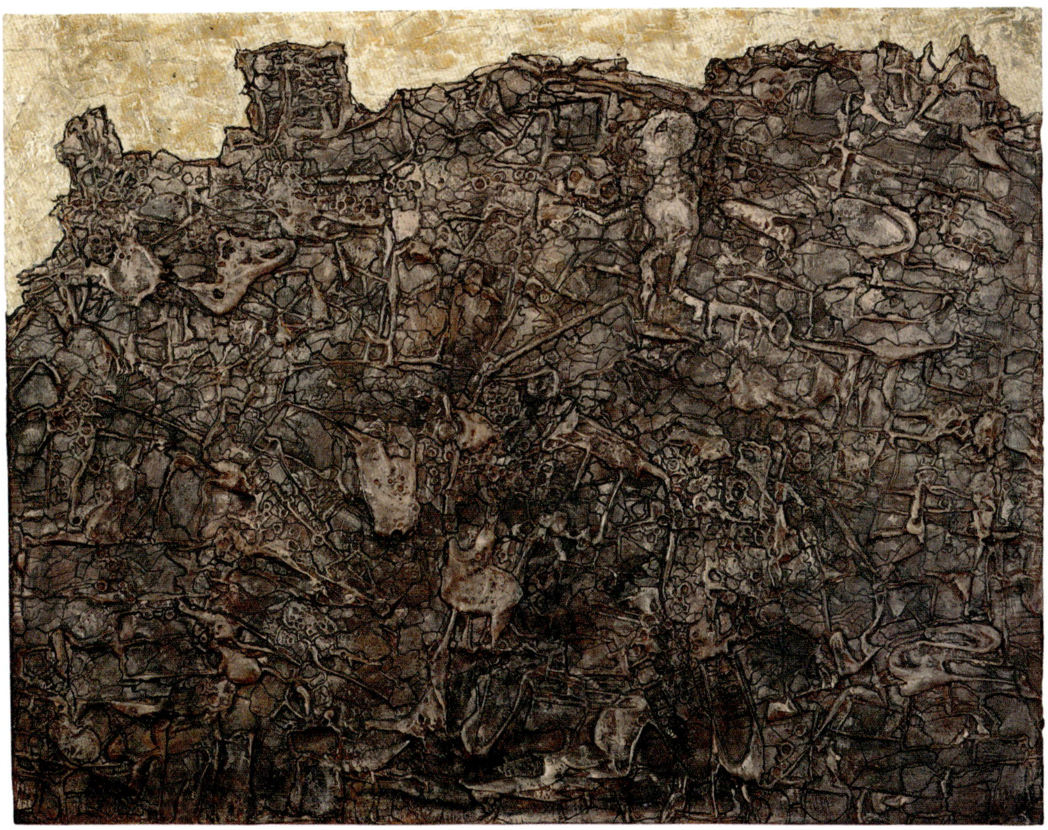

and prohibits with authority".[36] For him, buildings and city plans, both ancient and modern, were a potent expression of his much broader definition of architecture: "It is in the form of the Cathedrals and the Palaces that the Church and State . . . impose silence upon the masses,"[37] he wrote in his 'Dictionnaire' section of the Surrealist journal *Documents* in 1929. Indeed, 'form' itself was the corollary of architecture in his worldview. Bataille's definition of *informe*, in another edition of *Documents*, provided a destructive counter to this architectural hegemony. *Informe* was that which had the power to de-categorise and dismantle systems of authority and control; "to affirm on the contrary that the universe resembles nothing and is nothing but formless", he wrote, "comes to the same thing as saying that the universe is something like a spider or gob of spit".[38] For artists like Dubuffet and Fautrier, finding ways to express something like the *informe* became central to their projects. In post-war Paris, Bataille's earlier definition of 'architecture' and, by implication, 'form' seemed like a prescient insight into Fascism. Therefore, by manifesting its opposition in the formlessness of its base materials, such as mud, sand and pigment, 'matter painting' struck a blow for individual liberty whilst registering the painful scars of war. It also made a statement that impacted upon the aesthetics of formalist Modernism, particularly of the sort that had its architectural champion in Le Corbusier in Paris and, of course, was being implemented in grid-pattern tower-blocks across London.[39]

Auerbach's rendering of the architectural reconstruction of London as a struggle for "formal grandeur" from within the inherent formlessness of paint speaks to this discourse at a fundamental level. In Auerbach's building-site paintings formlessness is made manifest in the thick, wrinkled and puckered surface of his paint, defying visual comprehension as anything but 'matter'. But this happens only as a necessary condition of his achievement of a powerful formal composition. It was this structured side of Auerbach that Sylvester emphasised in his most considered and penetrating article about the artist's work from this period, entitled 'Nameless Structures', which accompanied his 1961 Beaux Arts Gallery exhibition.[40] In it he describes the way in which Auerbach's paintings express "a degree of architectural resolution that we normally associate with kinds of art that are concerned with constructing ideal forms; only we feel these forms have been arrived at empirically . . . . These structures seem to be known from the inside, as if the painter had become each object, had become the space, had become the light, and had painted them from the inside out."[41] The structures are, for him, "nameless" because they are reality re-made rather than representations of pre-existing named objects. Sylvester was able to see beyond the formless aspects of Auerbach's paintings, but did not fully account for the insistent presence of paint as an intractable substance in his work. Formlessness in Auerbach's paintings is more than a "by-product", as Sylvester had it, but is certainly not an end in itself, as might be said of Fautrier and Dubuffet.

Recasting Sylvester slightly, formlessness can be seen in the paintings as that which allows form to exist even as it works to deny that very existence. It is life in a hostile environment, and it reminds us of Auerbach's memory of seeing on the building sites "survivors scurrying among a ruined city".

It seems likely that Auerbach's personal history and situation, as a Jewish orphan in a devastated foreign city inching its way back to some semblance of normality, gave urgency to his desire to create a new unity of form in his paintings, holding formlessness, and perhaps the destructiveness of what Sigmund Freud called the 'death drive', at bay.[42] Auerbach's apparent desire, recognised by Sylvester, to inhabit his building sites by repeatedly painting and repainting them to create new form might also be thought of in the psychoanalytic terms propounded by Melanie Klein, both before and after the war. Her ideas were being introduced into the realm of art and architecture in Britain in a series of books by Adrian Stokes published during this post-war period, although Stokes was principally part of a very different set of British-based artists.[43] Klein's emphasis upon an individual's repeated efforts to enact reparation upon the destructive and fragmenting effects of past trauma finds interesting echoes in Auerbach's building-site works. Not only are the paintings products of Auerbach's painterly oscillation between destruction and re-formation, but the rebuilding of London itself must have seemed like an endless process of the same impulses set in train across the city. Klein's work is interesting, however, not as a psychoanalytic reading of Auerbach as an individual but as an element of the intellectual climate of post-war Britain within which his paintings were received. Her work also offers a way of understanding the paintings as a response to the trauma of war on a broader social and cultural level. Sylvester's writing on Auerbach, for example, clearly bears the trace of such thinking, and he was an admirer and friend of Stokes. It is pertinent that in both Klein's and, particularly, Stokes's writings it is an architectural metaphor which is used prominently to convey the theme of reparation, in ways that work against the expression of the *informe* in France and elsewhere. Unequivocally, for Stokes, it is the drive towards a unifying architectural form that is art's highest purpose. In his chapter from *Smooth and Rough*, 1951, entitled 'The Sense of Rebirth', he maintains that "fine building exemplifies the reparative function of art".[44]

Bataille's notion of the *informe* perhaps seemed more resonant as a liberating force in a city, like Paris, which had been occupied but kept most of its built form intact, than it did in London in the immediate post-war years. The links between Bacon's work and Bataille's thought are the major exception to this, and arguably not the only one. Indeed, Auerbach's repeated struggle for a profound strength of formal design in his paintings of London's rebuilding was certainly not an 'architectural' statement of authority in the Bataillian sense; nor was it the brightly progressive Modernist rhetoric of post-war reconstruction. But neither was it the unifying expression of the reparative impulse in quite the way the Stokes envisaged artistic ideals. The paintings' making appears too 'hard-won' (to recall Richard Morphet's phrase) for all this,[45] and their imagery and handling consistently transgress themes of order and control, oscillating between creation and destruction. Auerbach has previously borrowed Robert Frost's phrase about poetry to describe his own ambition: "A great painting is like ice on a stove. Its shape riding on its own melting into matter and space, it never stops moving backwards and forwards."[46]

Auerbach's building-site paintings do not sit comfortably within these various visions for art in the post-war world, but this is not to suggest that they stand aloof from the artistic discourses that animated the period. The works engage directly and revealingly with central concerns of art in Britain, Europe and indeed America during the 1950s and early 1960s. As James Hyman has discussed in his book *The Battle for Realism*, this was a period of intense debate as to the direction art was taking in Britain, with the role of representation being hotly contested across the art-political spectrum.[47] Berger and Sylvester emerged as leading figures, often in opposition to one another, and it is revealing that they should engage in different ways with Auerbach from the outset. Their repeated attempts to gain critical purchase on his paintings demonstrate not only the importance of Auerbach's work but also its rebarbative character. It was this that appealed particularly to Sylvester, as he put it in 1954: "His conception of painting is so extreme as to be almost absurd [yet his works] reveal qualities that make for greatness in a painter."[48]

The Building Sites and the Sublime

Ultimately Auerbach's building-site paintings speak of the repeated pushing and testing of the limits of form and structure. When American Abstract Expressionism arrived in London at the Tate Gallery's 1956 exhibition, *Modern Art in the United States*, Homer Cahill's introductory catalogue essay suggested, "One might say that they have gone back to beginnings, to 'the first division of chaos at the origin of painting'".[49] It was precisely this that Auerbach had been inspired to enact by London's building sites several years earlier. Willem de Kooning's work in particular (see fig. 11) was a source of interest for Auerbach, because it chimed strongly with his own concerns for pushing formal limits in both the figure and cityscape.[50] Auerbach's somewhat looser and more gestural paintings later in the series, such as *Shell Building Site: from the Thames* (cat. 9) and *Rebuilding the Empire Cinema* (cat. 13), owe something to the experience of seeing de Kooning's work as well as to his admiration of Chaim Soutine. But they are firmly rooted in Auerbach's slowly developing and intense exploration of his subject-matter, and their apparent

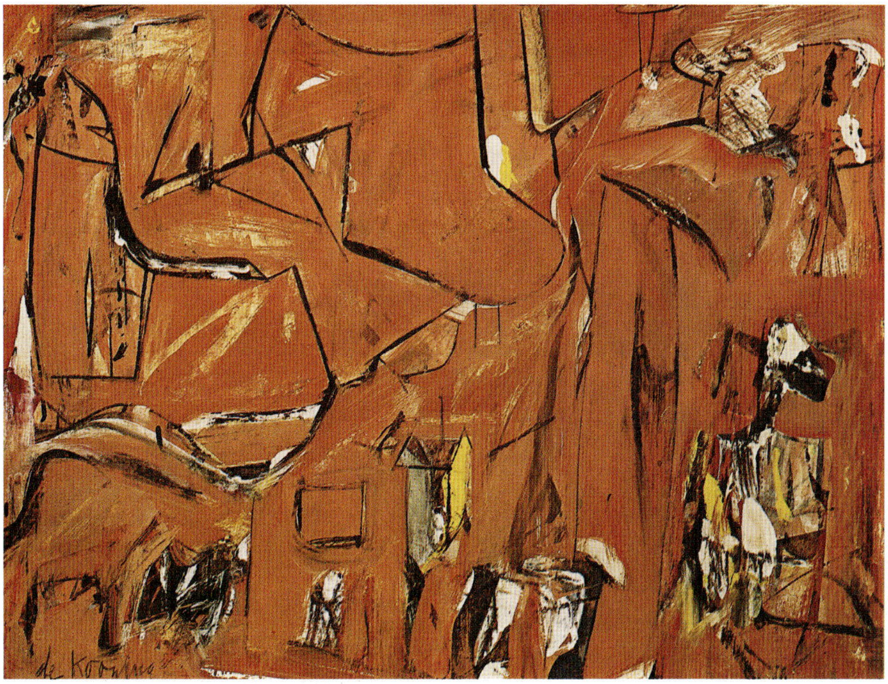

Fig. 11
Willem de Kooning,
Ganesvoort Street (New York)
1949–51
Oil on board, 76 × 101.5 cm
Collection of Harry W. and
Mary Margaret Anderson

spontaneity remained the result of months of repeated rehearsal in the studio. However, these later works do mark a development in the series. The darkness and density of his earlier building-site paintings, such as *Building Site, Earls Court Road* and *Building Site near St Paul's* (cat. 3 and 6), speak partly of the trauma of the early post-war years, and implicitly question the possibility of finding a pictorial language to express a city deranged by war. The later paintings powerfully express the excitement and awesome spectacle of the sites. In doing so they can be seen as an evocation of the sublime, which past artists had found in the mountains but which Auerbach found in the bombed and excavated "precipice and mountain and crags" of post-war London.

Certain 'Modern Masters' may have provided succour, but Auerbach's exploration of London's building sites was entwined at a more profound level with his experience of the Old Master and Romantic painters in the National Gallery. In a surviving sketchbook page from this period we find a sketch of a building site on the verso of a study after Turner's *The Parting of Hero and Leander*, which makes the connection explicit (figs. 12 and 13). Undoubtedly there would have been many other comparable studies from different paintings and building sites in sketchbooks that he has destroyed. Turner's painting conjoined the motifs and painterly effects that contemporary viewers, when it was first exhibited at the Royal Academy of Art in 1837, would have associated with an expression of the sublime. Dramatic storms breaking over a turbulent sea are set amidst a treacherous-looking mountain landscape. A fantastical city (based upon the study of classical ruins) is stacked precariously upon the rocks, stretching up into the hazy mists of the sky. This all speaks of the sublime, ineffable experience of overwhelming terror and excitement that gripped the later eighteenth- and early nineteenth-century imagination. The energy of Auerbach's study suggests that it also gripped him. It is clear that he has understood the painting to be a battle between the structuring geometric forms of the architecture and the wild destructive forces of the storm-possessed elements. Man's unwinnable intellectual and artistic contest with the incomprehensible scale and energy of nature was a Burkean and Kantian understanding of the sublime which not only energised the Romantic vision: its effects can be felt throughout the twentieth century as well, not least in the 'form' and 'formless' debates discussed above and, powerfully, in Auerbach's most vertiginous paintings, such as *Shell Building Site: from the Thames* and *Rebuilding the Empire Cinema*, among others.

For Auerbach, the building sites were a contemporary equivalent of a sublime landscape – one that could inspire the fear, excitement and strangeness of an uncharted mountain terrain. He recalls (see cat. 8) that some, such as the earthworks of the Shell Centre construction, were so powerfully composed that they could have been put in a museum without artistic intervention. However, Auerbach's paintings are an evocation of his experience of the sites recast as his own battle for expression in paint rather than an attempt to capture a sublimely picturesque view. One of the things that impressed him about artists such as Turner and Poussin was the way in which they could conjure a city of formal complexity and grandeur from their imaginations – "a marvellous faculty for building

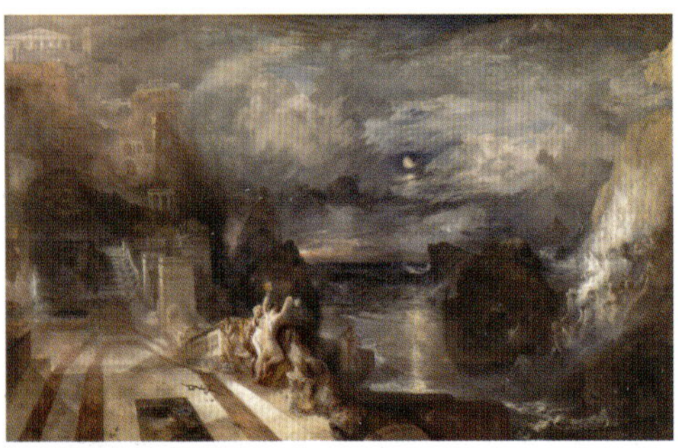

Fig. 12
J.M.W. Turner, *The Parting of Hero and Leander*
before 1837
Oil on canvas, 146 × 236 cm
The National Gallery, London

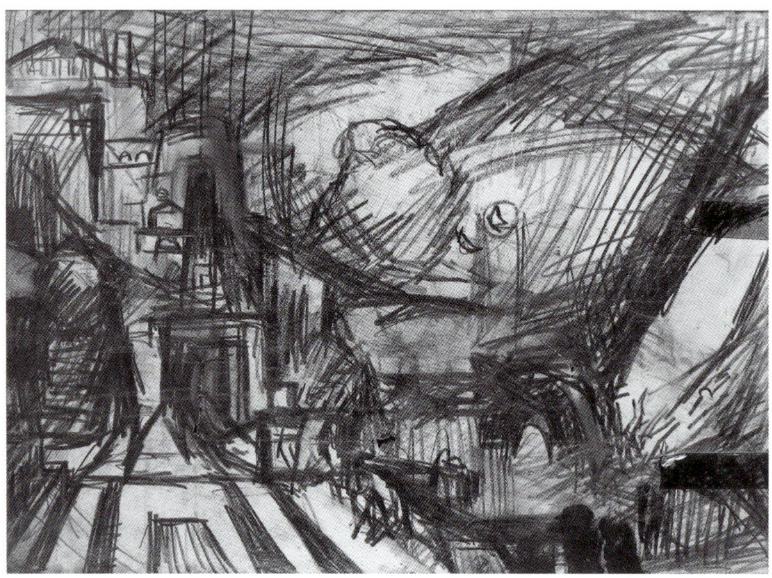

Fig. 13 RECTO

ideal architectures on the top of hills in the distance".[51] These 'architectures of the mind' are rarely stable affairs and are either pitted against the forces of nature or else – being based upon classical ruins – inherently foretell their own destruction. Bomberg had shown Auerbach Piranesi's *Carceri* series (see fig. 50) during his early days as an art student and they made a lasting impression, which can be seen surfacing particularly in paintings such as *Building Site near St Paul's*, *Oxford Street Building Site I* and *II* and *Rebuilding the Empire Cinema* (cat. 6, 11a, 11b and 13). Piranesi's imaginary 'prisons' express vividly, through form and composition alone, the obsessive condition of man's (and artists') overwhelming desire for control in the face of disorder.

Auerbach's own work might be considered as repeated attempts to construct pictorial architectures from the chaos which preceded architecture that he found on the building sites. Far from being freely expressive and wildly gestural evocations of his subject-matter, the paintings are underwritten by a strong sense of control and discipline. His constant reworking of the image with close reference to working drawings is one side of this. Another is his use of a restricted palette and his rarely wavering commitment to constructing his composition through an economy of straight lines – this is what makes the curling rope in the two Oxford Street paintings such a noticeable element (cat. 11a and 11b). His first dealer Helen Lessore put it very well when she wrote: "Part of his procedure (making the cage) involves an arbitrary limitation of means. In this he is like a poet who wants to purify and revivify a language, and so deliberately deprives himself [of] the use of a great many words [and] tries to wring the last drop of their possibilities from them."[52] And yet the

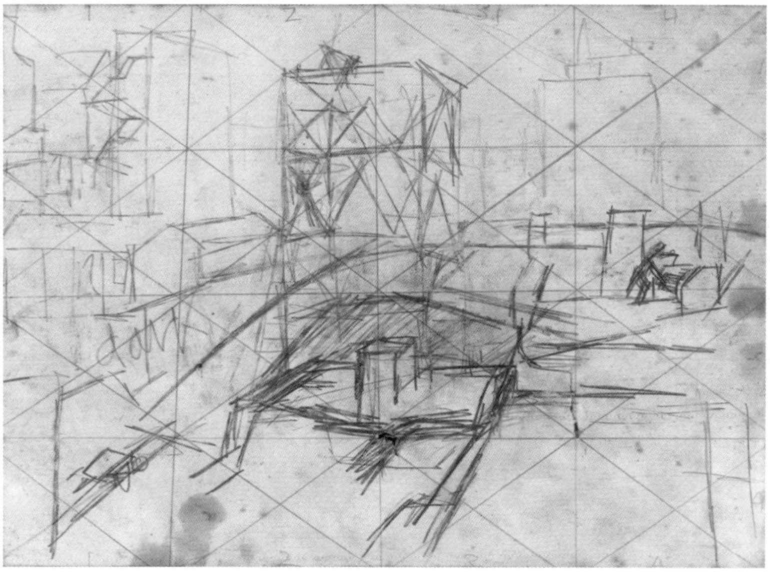

Fig. 13 VERSO

Fig. 13
Frank Auerbach, *Study after Turner's The Parting of Hero and Leander* (recto); *Sketch of a Building Site* (verso)
c. 1952
Pencil on paper, 25.1 x 34.9 cm
Private collection

ultimate impossibility of securing an irreducible image or an ideal architecture is somehow registered in the paintings, even as Auerbach achieves a formal grandeur. As he put it in 1959, "One never has power over anything, can never do anything clearly or purely . . . . One tries to do what one can't do."[53] Quite how this paradox operates in the paintings defies convincing explanation: it is to be experienced not written about, and perhaps it is this that constitutes their expression of a modern sublime.

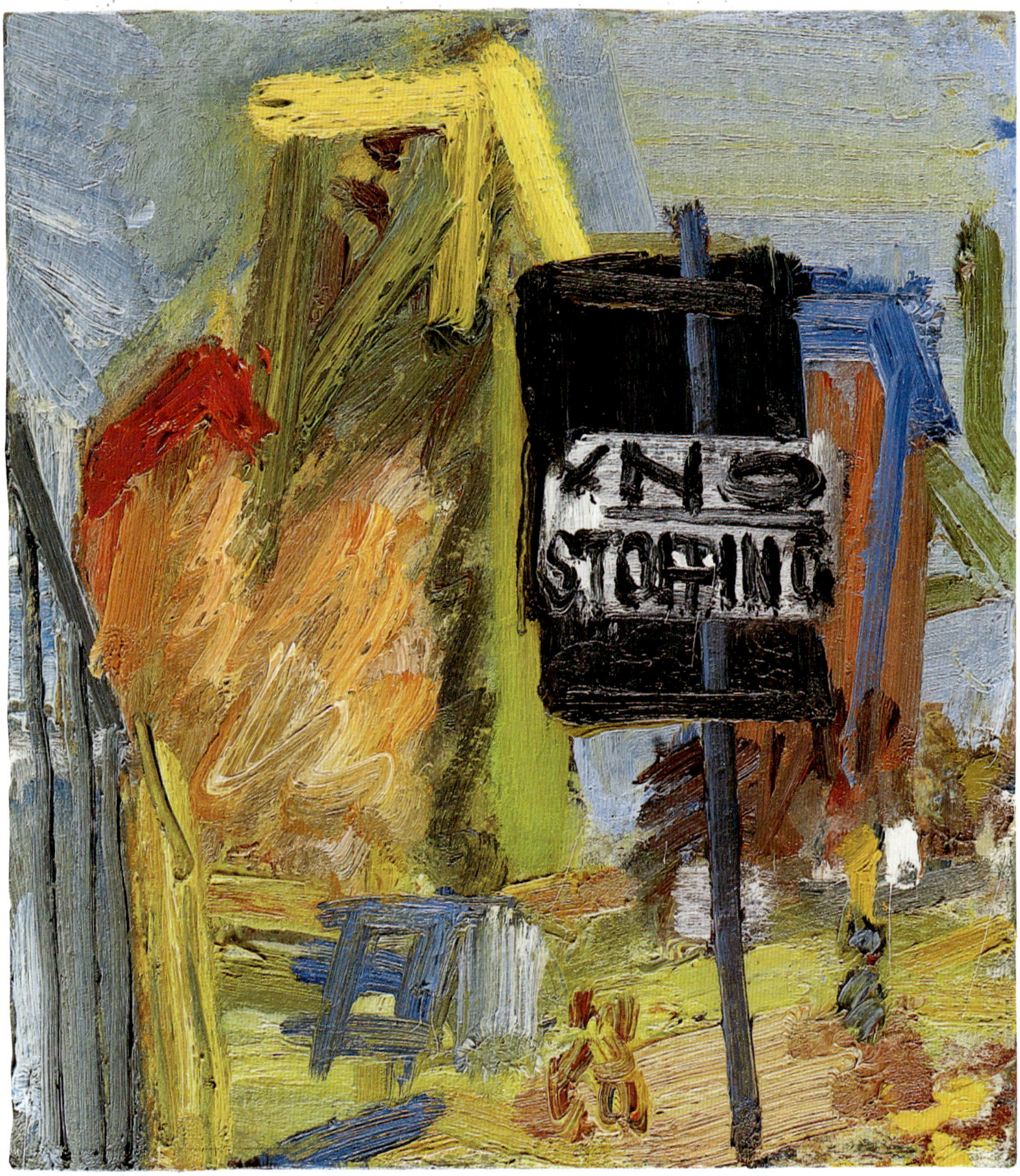

Fig. 14
Frank Auerbach, *Tower Blocks, Hampstead Road*
2007–08
Oil on board, 50.8 × 45.7 cm
Courtesy of Marlborough Fine Art, London

Conclusion – Life and Death and Rebirth in the Modern City

The writer Bryher (quoted above) was astute in observing that the War had driven people back to their beginnings. The art historian Lyndsey Stonebridge has suggested that a number of writers and artists in Britain during and immediately after the Blitz expressed a sense that "war is birth the wrong way round"[54] – death and destruction a condition of life and creation. For Auerbach the building sites of London seem to have been a powerful encapsulation of this disconcerting post-war phenomenon, in which personal history, collective experience and the foundations of art were caught up. As such they speak of a particular time and place and need to be seen as a profound part of London's post-war history. That the paintings resist straightforward assimilation within the art and ideas of the period (as this essay has tried to show) is a sign of their originality. More importantly it is a demonstration of the fact that for Auerbach the unprecedented spectacle of the building sites demanded a new language of painting. *Rebuilding the Empire Cinema* (cat. 13) would prove to be his final building-site painting. After this he largely confined his landscape and cityscape subjects to the proximity of his Camden Town studio, and has never retuned to construction sites as a theme. However, the building-site series would provide important foundations for his future work and its legacy as a battle for creation fought in the teeth of destruction can be seen in his paintings to this day (see fig. 14).

The relevance of the series stretches beyond its immediate post-war context and Auerbach's future work. It is revealing that Rembrandt's small painting in the National Gallery, *The Lamentation* (see fig. 41), was a constant source of inspiration to Auerbach during the period. Auerbach's and Rembrandt's shared concern to find a way of embodying in paint the liminal period between death and resurrection, unites these two artists across the centuries. As meditations on the theme of destruction and creation Auerbach's building-site paintings speak to broader concerns of art and culture over a much longer history.

Looking back recently at the paintings after some fifty years, Auerbach mused: "If I think about them now, long after, there's just a tiny echo of the first book of Genesis in all these building sites".[55]

Notes

I am indebted to Margaret Garlake, Christopher Green and Ernst Vegelin van Claerbergen for their invaluable comments on earlier drafts of this essay.

1 Frank Auerbach, 'Fragments from a Conversation', in David Wright and Patrick Swift (eds.), *X: A Quarterly Review*, vol. 1, no. 1, London (Barrie and Rockliff) 1959, p. 34.

2 Lionel Esher, *A Broken Wave: The Rebuilding of England, 1940–1980*, London (Allen Lane) 1981, p. 72.

3 The term 'creative destruction' has been used in a number of contexts but is best known as a concept in economics. The phrase was used by Joseph Schumpeter in 1942 to describe the process of radical innovation within capitalist economies, such as the introduction of a new product to the market which quickly rendered earlier alternatives obsolete, or the improved competitiveness of a company which would drive others out of the market. Economic forces of creative destruction undoubtedly propelled the post-war rebuilding of London, most explicitly in the office building boom and the rise of the commercial developer: Joseph Schumpeter, *Capitalism, Socialism and Democracy*, London and New York (Harper) 1942.

4 Major accounts of Auerbach's biography and work can be found in Robert Hughes, *Frank Auerbach*, London (Thames & Hudson) 1990; Catherine Lampert (ed.), *Frank Auerbach: Paintings and Drawings 1954–2001* (exhibition catalogue), Royal Academy of Art, London, 2001; and William Feaver, *Frank Auerbach*, New York (Rizzoli) 2009 (forthcoming). I am grateful to William Feaver for sharing his text with me in advance of publication.

5 Auerbach/Wright interview 2009.

6 *Ibid.*

7 J.M. Richards, *The Bombed Buildings of Britain: A Record of Architectural Casualties, 1940–41*, London (Architectural Press) 1942, p. 2.

8 Richards claimed that a bombed ruin possesses "an aesthetic peculiar to itself [and] contrives its effects out of its own range of raw materials ... the scarified surface of blasted walls, the chalky substance of calcined masonry, the surprising sagging contours of once rigid girders and the clear siena colouring of burnt-out brick buildings, their rugged cross-walls receding plane by plane, on sunny mornings in the City": *ibid*, p. 3.

9 Virginia Button, 'The Art of the Second World War', in Chris Stevens (ed.), *The History of British Art, 1870 – Now*, London (Tate), 2008, p. 118. The extent to which these neo-Romantic and neo-Picturesque works represented an all too consolatory account of the British experience of war, compromised by State sponsorship, has been much debated. The works produced under the WAAC scheme are inevitably more open to such claims than others. In the hands of an artist such as John Minton (who would later become one of Auerbach's tutors), the neo-Romantic idiom could be turned to account for the personal trauma of war rather than engendering a collective spirit of stoicism and renewal, as in his *Self-portrait with Ruins*, 1941, Imperial War Museum. Much of Sutherland's and Piper's wartime work (especially that not produced for the WAAC) can also be seen as bearing witness to anguish, loss and displacement.

10 John Piper, 'Pleasing Decay', in *Buildings and Prospects*, London (Architectural Press) 1948, p. 99. Piper's argument had been explicitly prefigured in Sutherland's drawing, *Devastation: Composition*, 1942 (private collection), which transposes Picasso's *Guernica* on to a Blitz scene.

11 Auerbach 1959, p. 32.

12 Auerbach/Wright interview 2009.

13 Hughes 1990, p. 32.

14 Auerbach/Wright interview 2009.

15 *Ibid.*

16 'Pictures on Exhibit', February 1956 (publication unknown), cutting held in the Beaux Arts Gallery Archive.

17 David Sylvester, 'Young English Painting', *The Listener*, 12 January 1956, p. 64.

18 There are some examples by other artists. John Berger made a number of paintings of the construction of the Festival of Britain on the South Bank in 1950 and David Tindle painted the cranes on the same site a decade later as the Shell Building went up there. But theirs were not major artistic statements, nor was the subject a sustained source of inspiration for either artist. London building-sites were a theme in Prunella Clough's work of the 1950s, along with other subjects of industry and manual labour, which fed her powerfully idiosyncratic attempts to create synthesis between realism and abstraction. But, with few exceptions, London's building sites belonged to Auerbach and Kossoff.

19 Auerbach 1959, p. 32.

20 Peter Ackroyd, *London: The Biography*, London (Chatto and Windus) 2000, p. 745.

21 Bryher (Annie Winifred Ellerman), *The Days of Mars: A Memoir 1940–1946*, London (Calder & Boyars) 1972, p. 150. This quotation is cited from Lyndsey Stonebridge, 'Bombs, Birth and Trauma: Henry Moore's and D.W. Winnicott's Prehistory Figments', in Jane Beckett and Fiona Russell (eds.) *Henry Moore Critical Essays*, Aldershot (Ashgate) 2003.

22 Auerbach recalls that during his wartime stays in London for part of the school holidays he had heard air raid sirens and bombs dropping, including, on one occasion, close to where he was standing. However, he did not find any of these experiences traumatic: "I think it was fairly routine for people *living* in London, and it never seemed momentous to me": letter from Frank Auerbach to Barnaby Wright, 18 April 2009.

23 Auerbach/Wright interview 2009.

24 Auerbach 1959, p. 31.

25 Auerbach/Wright interview 2009.
26 Neville Wallis, *Observer*, 29 November 1959. The analogy could be stretched further as Auerbach's epic and lengthy process of painting came close to matching the time-scales involved in the buildings themselves. For example, by the time he had finished his painting of the early stages of construction of the Time-Life offices on Bruton Street (cat. 2) the building had been completed and opened for business.
27 Auerbach/Wright interview 2009.
28 Stephen Bone, *Guardian*, 11 January 1956.
29 Rosalind E. Krauss, 'Grids', in *The Originality of the Avant-Garde and Other Modernist Myths*, Cambridge, Mass., and London (MIT Press) 1986, p. 11.
30 See Margaret Garlake's essay in this catalogue.
31 John Berger, *New Statesman and Nation*, 14 January 1956.
32 David Sylvester, 'Nameless Structures', *New Statesman*, 21 April 1961.
33 For a discussion of these debates see Caroline A. Jones, 'Form and Formless', in Amelia Jones (ed.), *A Companion to Contemporary Art Since 1945*, Oxford (Blackwell) 2006, pp. 127–44.
34 Dawn Ades, 'Web of Images' in *Francis Bacon*, (exhibition catalogue), Tate Gallery, London, 1985.
35 See Paul Moorhouse's essay in this catalogue for a discussion of existentialism in relation to Auerbach.
36 Georges Bataille, 'Architecture', in *Documents*, no. 2, Paris 1929, p. 117.
37 *Ibid.*
38 Georges Bataille, 'Informe', in *Documents*, no. 7, Paris 1929, p. 382.
39 The relationship between the work of artists associated with *L'Art Informel* and the Modernism of Le Corbusier and his followers is far from straightforward and was by no means one of opposition. For example, Le Corbusier's much imitated use of rough concrete in his post-war work makes a point of connection with the aesthetics of 'matter painting'. This would feed into the architectural theories of 'Brutalism' from the 1950s onwards in England and elsewhere.
40 Sylvester 1961. Although a review of Auerbach's 1961 exhibition, Sylvester's piece was intended as a summation of his career to date.
41 *Ibid.*
42 Freud's understanding of the 'death drive' is interesting in this context, as he discusses its destructive impetus as operating in interaction with the opposing 'life' instinct towards cohesion and unity: Sigmund Freud, 'Beyond the Pleasure Principle' (1920), in Angela Richards (ed.), *On Metapsychology: The Theory of Psychoanalysis* (The Pelican Freud Library), London (Pelican) 1984.
43 Stokes was principally associated with artists such as Ben Nicholson and Barbara Hepworth who had moved to St Ives in Cornwall shortly before the Second World War, as did Stokes. His books from this period include *Art and Science*, London (Faber & Faber) 1949, *Colour and Form*, London (Faber & Faber) 1937 (revised 1950), and *Smooth and Rough*, London (Faber & Faber) 1951. I am indebted to Christopher Green's discussion of Klein and Stokes in relation to architecture in Christopher Green, *Picasso: Architecture and Vertigo*, New Haven and London (Yale) 2005, pp. 34–38.
44 Stokes 1951, p. 57.
45 Richard Morphet (ed.), *The Hard-Won Image: Traditional Method and Subject in Recent British Art* (exhibition catalogue), Tate, London, 1984.
46 Catherine Lampert (ed.), *Frank Auerbach* (exhibition catalogue), Arts Council of Great Britain, Hayward Gallery, London, and Fruit Market Gallery, Edinburgh, 1978, p. 20.
47 James Hyman, *The Battle for Realism: Figurative Art in Britain During the Cold War, 1945–1960*, New Haven and London (Yale) 2001.
48 David Sylvester, 'Round the London Galleries', *The Listener*, 12 August 1954.
49 Homer Cahill, 'Abstract Expressionism', in *Modern Art in the United States: A Selection from the Collections of the Museum of Modern Art, New York* (exhibition catalogue), Tate Gallery, London, 1956, p. 24.
50 De Kooning was represented at the exhibition by *Painting*, 1948; *Woman I*, 1950–52, and *Gansevoort Street*, 1949–51 (here fig. 11).
51 Auerbach/Wright interview 2009.
52 Helen Lessore, *A Partial Testament*, London (Tate) 1986, p. 59.
53 Auerbach 1959, p. 32.
54 Stonebridge 2003, p. 111.
55 Auerbach/Wright interview 2009.

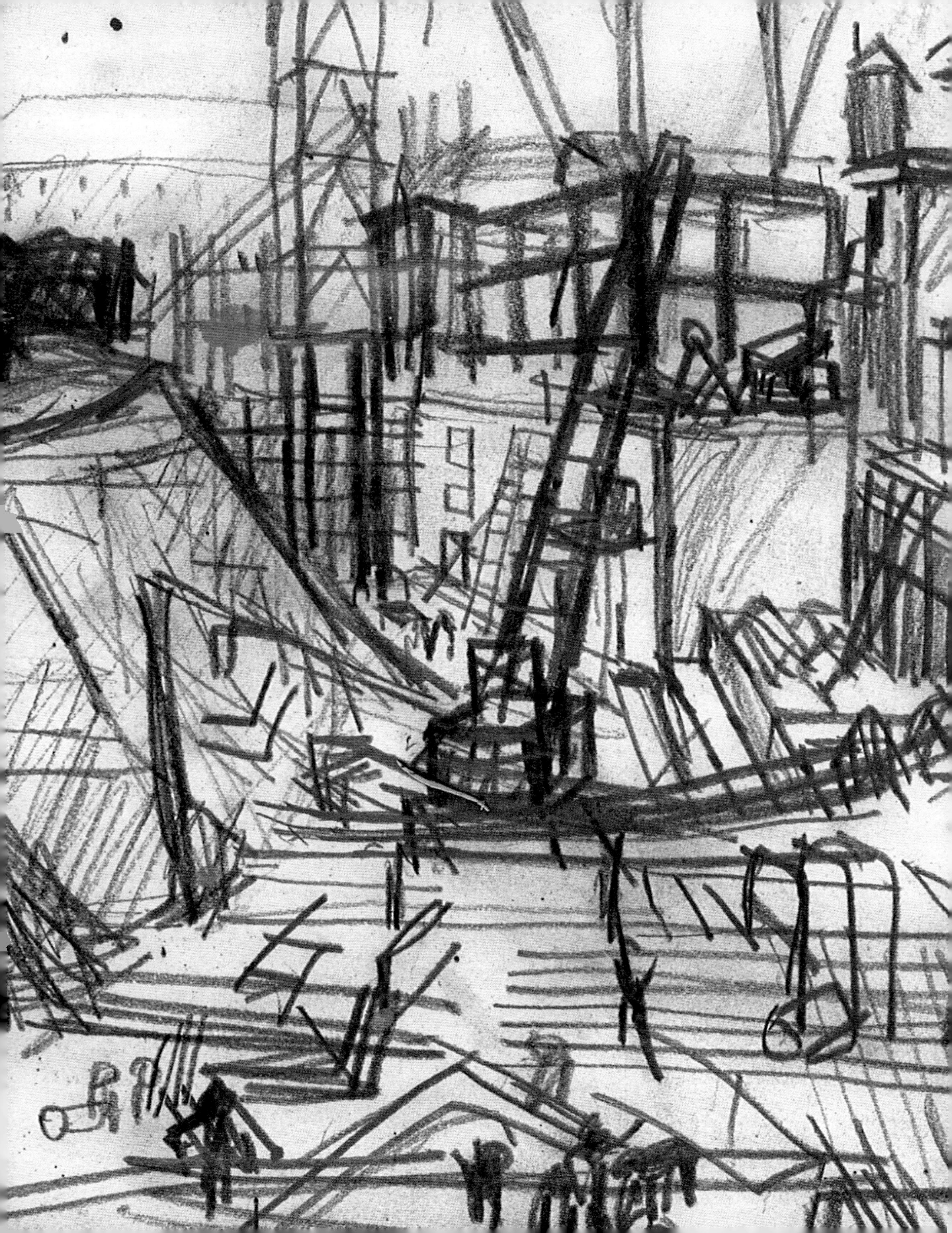

Prefabs and Stubby Skyscrapers: Rebuilding London 1945–60

MARGARET GARLAKE

"Now the place where the bomb had fallen .... About three acres lay flat, bared of building. Almost – it was a half-job – the place had been neither cleared, nor left .... All the loose rubble had gone, or been piled up against walls, or the fence; but pits of water marked old basements, and sharp bits of wall jutted, and a heap of girders rusted. The ground floor of a house stood, shacked over with iron, in the middle, and a single wall reared high up from it, intact, with fireplaces one above another."[1]

The post-war London in which Frank Auerbach began his career has disappeared almost as completely as the medieval city, but those who remember it, however distantly, are left with images so strange, so ineradicable that they scarcely translate into the terms of the everyday world.[2] That every aspect of life involved bringing order to chaos is vividly underlined by Auerbach's building-site paintings, dominated by geometric structures imposed on the gaping holes in the ground first made by bombs, then reconfigured by construction workers in a process that echoed his own practice of erasure and reinstatement. Early post-war London was degraded, dirty, makeshift; bombed buildings were 'made safe' then, waiting their turn in the long rebuilding queue, became the adventure playgrounds of their day.

'Post-war' describes as much a cultural process as a chronological period, a set of circumstances lived through and turned to account. The range of painters' responses to the visual stimulus of the city in the decade after 1945 is as good an indicator as any of the many ways in which those years were experienced and understood. Though the National Emergency was declared over in 1948, the travails of early peacetime were lengthy, extending far beyond the building trades. The Festival of Britain in 1951 encouraged perceptions of improvement and a further turning-point came in November 1954, with the end of building controls. For the purpose of this essay the cancellation of restrictions and taxes on new buildings can be taken as the end of the post-war period.

This essay is about the rebuilding of London, an undertaking all too often disconnected from architecture or aesthetic and historical considerations. Planners, builders and architects were required to negotiate not only the social aspirations to which the Beveridge Report had been both trigger and response, but also memories of the Depression, a near-bankrupt economy, a national failure to do the sums, industrial instability, the far more terrifying instability of East-West relations – a fluid set of unpredictable circumstances. The fundamental problem was economic, resulting in a dilemma over "the relative degree of priority which can be given to building in this country in relation to the needs of industry and export and to the situation in Europe ... the relative importance of housing and education as well as [to] the reconstruction of industry for peace-time needs".[3]

OPPOSITE
Detail cat. 8 (iii)

Destruction – Planning – Legislation

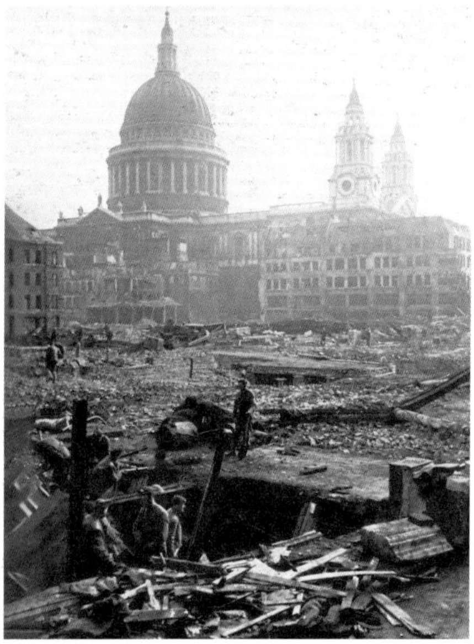

Fig. 15
The bomb-damaged ruins of Paternoster Square with St Paul's Cathedral in the background, from C.H. Holden and W.G. Holford, *The City of London: A Record of Destruction and Survival*, London (Architectural Press), 1951

The Blitz began on 7 September 1940 and lasted for 76 nights, directed first at the East End, the Dockland area which was already beset by "obsolescence, overcrowding, insanitary conditions, lack of open spaces, inadequate road systems".[4] By mid October 16,000 houses had been destroyed, nearly 200,000 had been damaged, 250,000 people were homeless and over 11,000 had been killed. On the night of 29–30 December the Guildhall and eight Wren churches were destroyed. Overall one third of the City was demolished, especially in the vicinity of St Paul's (figs. 15 and 16). The focus of the bombing then shifted to central London, where the West End, the retail and entertainment centre, suffered badly. Elsewhere bomb damage was lighter or patchy, so that it was not unusual for one or two houses in a Victorian terrace to be destroyed, leaving enduring evidence of the event in the form of an anachronistic infill.

Discussion of the priorities for reconstruction began in 1943, long before the last bombs had fallen or the final damage could be assessed. After D-Day active planning was in place for post-war housing and building. It was clear to all those professionally involved in reconstruction that planning was the key to solving London's long-terms problems of sub-standard housing, congested traffic, few open spaces and insufficient separation between residential and industrial areas. Eminent planning consultants were appointed and made recommendations for which they were highly praised, but in the event London was rebuilt piecemeal, as J.M. Richards, one of the editors of the *Architectural Review*, was to lament.[5] Awareness of the capital's problems had no doubt encouraged proposals, under consideration since 1935, to bring the whole of London under planning control. Private residential building had proliferated between the wars, encouraged by building societies and the speculative builders who created the 'suburban semi'.[6] At the same time the London County Council and the Borough Councils conducted vigorous housing campaigns, supported by the 1930 Housing Act, which provided subsidies for rehousing with a 'special subsidy for flats', which were traditionally unpopular and rare outside London and Liverpool.[7] However, all building programmes were superseded in October 1940 by the Government Building Programme, a huge project of "new construction for direct war purposes".[8]

A series of wartime reports reveals strenuous attempts to set the future parameters of planning. Among many schemes for reconstruction the most radical was the "futurist fantasy" proposed by the MARS Group in 1938, involving the demolition of the entire existing city.[9] More realistic proposals included a Beaux Arts scheme from the Royal Academy (denounced by the *Review* for its classical references) and a master plan (later described as "insipid") by the Royal Institute of British Architects' eclectic Reconstruction Committee, which also organised an exhibition, *Rebuilding Britain*, shown at the National Gallery in 1943.[10]

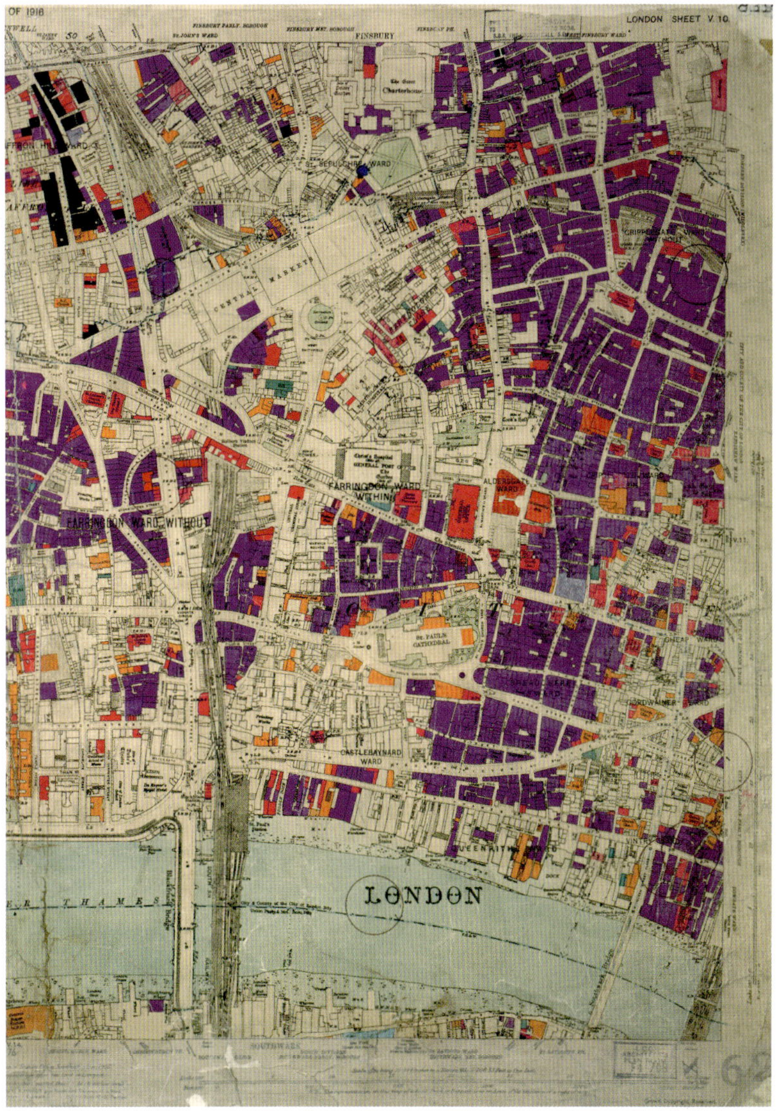

Fig. 16
A section of a bomb-damage map showing the area around St Paul's Cathedral, *c.* 1945. The purple-coloured areas indicate buildings which suffered "damage beyond repair". City of London, London Metropolitan Archives

The Barlow and Uthwatt reports (1940 and 1941) were fundamental: both informed the 1947 Town and Country Planning Act that "effectively set up the post-war statutory planning system".[11] Barlow (the Royal Commission on the Distribution of the Industrial Population) recommended the creation of a national planning authority and the redevelopment of congested urban areas for housing, with industrial development to be shifted evenly to the regions, while Uthwatt proposed that the Government be entitled to develop all land not already built on in return for compensation, with undeveloped and war-damaged sites in built-up areas being subject to compulsory purchase in order to house the homeless.

The much-acclaimed *County of London Plan*, 1943 ("the LCC plan") dealt specifically with London.[12] It adopted aspects of the Barlow report's recommendations on dispersal, though the authors were primarily concerned with the reconstruction of old London communities, a process that would "include the best of existing London ... enhance its strongly marked character ... respect its structure and spheres of activities, but at the same time, and drastically if need be ... remedy its defects".[13] Thus the authors made proposals for ring roads, utilities, parks, community centres and schools. They identified the south bank of the Thames between County Hall and London Bridge, with the sadly degraded West End, as important areas for redevelopment, though the East End took priority for housing. It included the main Redevelopment Area, divided into "neighbourhood units", each with "a mixture of low-density housing and high-density flats, with schools, a community centre, shops and business premises".[14] Flats were an inevitable concomitant of the Plan's recommended population densities: at a median 136 people per acre, two thirds of the population would live in flats.[15] In 1946 the Government increased the subsidy per flat for blocks of four or more storeys with lifts. By 1949 the LCC had built 50,000 dwellings in a mixture of four- to six-storey blocks of flats and houses. Overall, approximately 220,000 houses were built annually in England, four-fifths of them by local authorities, though many were architecturally undistinguished and very few used industrial construction techniques.[16] Two years later a new Conservative government announced fresh guidelines for local authority housing, with a reduction in living space to cut costs.

Discussion of planning, determined in practice by immediate assessments of urgency, turned initially on the opposing priorities of housing and industry. While historians have generally acknowledged the inevitability – and propriety – of a welfare-based society after the War, Correlli Barnett has taken a contrary view, attributing Britain's industrial decline through the latter half of the twentieth century in part to decisions on planning. This is not the place to argue the case for the welfare state, born out of the travails of war, but conversely it is difficult to argue against the evidence of a decline that "began in wartime British dreams, illusions and realities".[17] In 1943 the Official Committee on Post-War Internal Economic Problems estimated that a labour force of one and a quarter million men would be required within four years of the war's end to build and repair the desired number of houses. This would cost £670 million, out of "a national rate of saving of less than £800 m".[18] The debate continued towards a tacit acceptance of the priority of housing, aided by the assumption of Lord Reith, Minister of Works, that no new factories would be built for the first two years of peace since industry would adapt factories built for war production.[19] In 1945 Duncan Sandys pronounced "... first things must come first. The houses must go up and nothing must stand in their way."[20] By 1945 222,000 dwellings in London had been either destroyed or irreparably damaged and none had been built for six years. The final allocation of funds in January 1945 gave £56 million to housing in the first post-war year, against £14 million for trade and industry.[21] A year later the housing shortage was such that thousands of people were squatting in empty blocks of flats. Two years after the end of the war only 97,000 new homes had been completed against an aspiration for 100–200,000 in the first year of peace.[22]

Reconstruction was determined by planning, manpower and the supply of materials, and was beset by controls to protect scarce supplies and the orderly fulfilment of the Government's plans. Writing in 1947, the architect Richard Sheppard emphasised the need to reorganise the building industry: "... the conditions of work [are] thoroughly unsatisfactory", and recruitment was difficult.[23] "The supply of building materials is also inadequate, ill-adjusted and badly distributed," he wrote – a situation exacerbated by the interdependence of industries: construction was linked with coal mining, the docks, the gas supply, iron and steel, the railways, road haulage and shipping.[24] Coal was in desperately short supply, steel and iron were required for export and timber could not be imported because of the shortage of foreign currency; as a result the construction industry was in "continuous crisis" and was delaying "the physical and financial aspects of recovery".[25] The Attlee government continued the wartime ban on strikes, which did not prevent numerous unofficial strikes, mostly in the mines and docks, where they caused acute economic disruption.[26] The frequency of strikes in the construction industry rose steadily until, between 1953 and 1959, it was "the most strike-prone industry with an average of 120 strikes per year", a figure that more than doubled between 1960 and 1968.[27]

Despite these problems the prospect of resuming normal building operations encouraged the emergence of the developer, who was soon to introduce a new dynamic to reconstruction. Even before the end of the War developers had identified weak points in the regulations; the Government, urgently needing office space, intimated that developers with appropriate sites would receive building licenses "subject to an occupational lease for a Government department". Consequently speculators took care to earmark buildings and sites suitable for commercial redevelopment. A 'speciality market' emerged, in the form of the 'lessor' scheme, involving both owners wishing to sell or eligible for compensation and district surveyors (who were open to a degree of corruption): if a surveyor issued a Dangerous Structure Notice the owner was compelled to repair the building and a license was automatically provided.[28]

In October 1951 Harold MacMillan was appointed Minister of Housing and Local Government with a remit to fulfil the Tory pledge of 300,000 houses a year. In November 1954 MacMillan's "bonfire of controls" became law, abolishing licences, daylight controls for City high-rise buildings and a development tax that had, "in effect, amounted to 100 per cent".[29] A building boom followed, fuelled by an "insatiable" demand for new office space, which "created fortunes for developers" on the basis of contracts arranged in the space of a few days.[30] Late in 1964 the Labour government brought the boom to an end with the 'Brown

Fig. 17
J.D.M. Harvey, *The Festival of Britain*
1951
The National Archives, London

Ban', which restricted office building in the south-east and forbade the construction in London of any office over 2,500 sq. ft., thus considerably encouraging decentralisation.[31] By this time, planners had ceased to encourage development at all costs, conservation was becoming a matter of concern and much adverse comment was made on "curtain-walled elevations [that] gave the appearance of huge cubic goldfish bowls".[32]

Architects – Buildings – Developers

Reconstruction demanded technical innovation and imaginative planning. The *Architectural Review* had campaigned since 1944 for a "resurrection" of picturesque planning principles ("England's paramount contribution to the optics of planning"), claiming that it "could revolutionize our national contribution to architecture and town-planning by making possible our own regional development of the International Style" as the basis of a local, English modernism "rooted in a regional identity".[33] The planning and form of public housing were hotly debated, with Frederick Gibberd's mixed development of low-rise flats and terrace

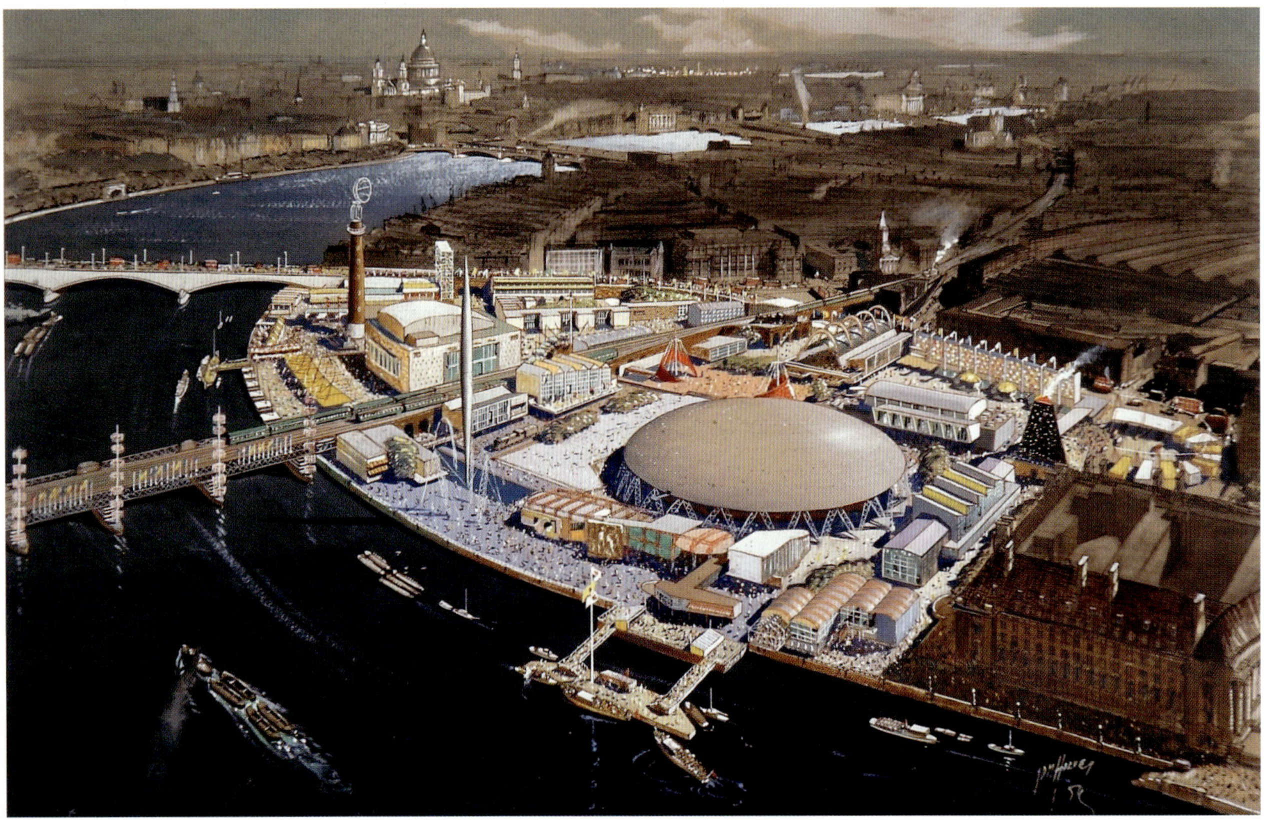

Fig. 18
Time–Life Building, Bond Street (Michael Rosenauer), 1951–53
Photograph: Conway Library, The Courtauld Institute of Art

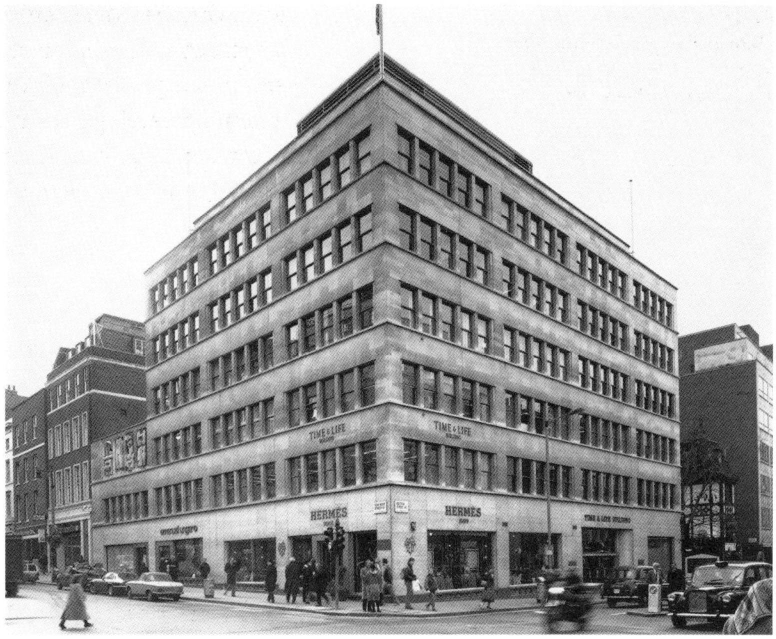

housing at Somerford Grove, Hackney (1947–49), much praised for its regional quality, demonstrated by a combination of modern flat roofs with traditional brick construction, while picturesque planning and visual principles flowered in the South Bank exhibition (1951), "where the Picturesque theory has been followed with triumphant results" (fig. 17).[34] One notably prestigious modern office-building to go up shortly after the Festival of Britain was the Time–Life Building on Bruton Street (Michael Rosenauer; fig. 18), which integrated the work of several artists and designers who had contributed to the 1951 exhibition; Auerbach produced a painting based on the early stages of its construction (cat. 2). Meanwhile the *Review* moved on to promote 'New Empiricism', developed by Swedish architects in an interrogation of the International Style and distinguished by its use of traditional materials and attention to detail and planning.[35] However, older 'principal architects' tended to be practical rather than aesthetically conscious, and accustomed to conventional construction methods finished in the conservative neo-Georgian style that was the centrist position for important public buildings.[36] This impasse was scarcely resolved before the late 1950s, when the long debate on historicist styles was finally displaced by advances in industrialised building techniques which by default produced a version of Modernist architecture across the capital.

Developers, inherently conventional, tended to employ older commercial firms ("the sort . . . that knew its way through the controls and did not waste time or money on 'architecture'") rather than younger architects.[37] After 1954 diversification took hold, with the establishment of new firms and the growth of commercial practices to accommodate an unprecedented amount of office building, but 1954 was also the year when Alison and Peter Smithson completed Hunstanton School in Norfolk. Informed by Mies van der Rohe's Illinois Institute of Technology campus, Hunstanton announced the conceptual approach to design that Reyner Banham named "the New Brutalism".[38] Grounded in Le Corbusier's Jaoul houses and the Marseilles Unité, it put paid to any lingering desire for the neo-Picturesque and offered an avant-garde alternative to the developers' glassy, curtain-walled modernity.

LCC – Planning – Monumentality

If, in today's terms, the planning process looks intolerably top-down, it is almost impossible to envisage an alternative approach, given the range of the problems and the level of docility recorded by surveys like Mass Observation.[39] To claim that there was "no doubt of the popularity of a thesis that argued for a greater level of public ownership and standardisation to overcome social inequalities" is to overlook the sheer inertia of a war-weary population unaccustomed to being asked its opinion, but it is impossible to over-emphasise the weight of aspiration invested in the modern individual home as a fitting marker of the new era of peace.[40]

As the principal provider of housing in the capital, it was for the LCC to develop a viable model. Prefabrication, made possible by the conversion

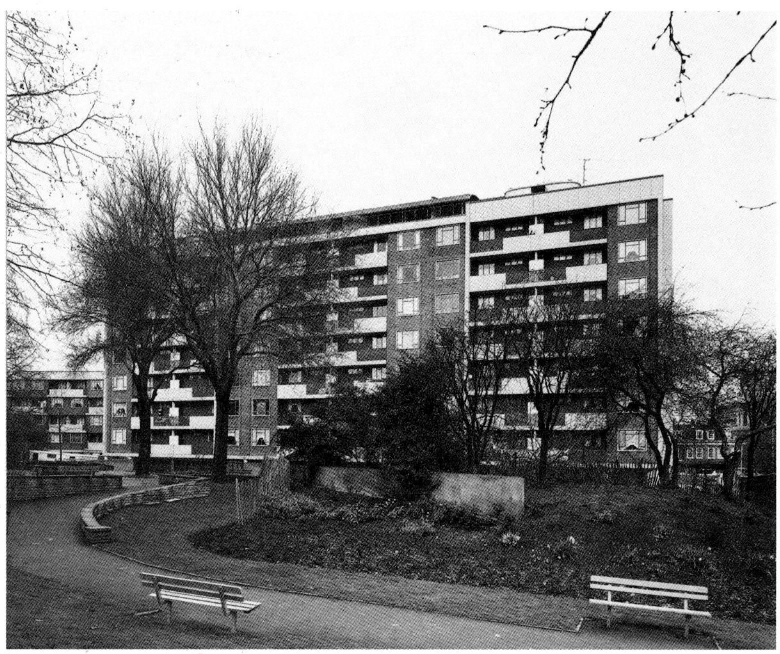

Fig. 19
Spa Green Estate,
Clerkenwell (Tecton), 1946–50
Photograph: Conway Library,
The Courtauld Institute of Art

of redundant aircraft factories, briefly appeared to offer a solution to the housing crisis: "Over 150,000 'temporary' Portal bungalows were run off in double-quick time".[41] Twice as many had been envisaged, but because of shortages of steel and timber the scheme was abandoned, though not before "almost every available bombed site, and even some public open spaces [were] now occupied by prefabs".[42] Though the prefab was identified as significant for its influence on the design of a variety of small houses, large-scale industrial construction, scarcely envisaged in 1946, was to feed the office boom rather than solve the housing crisis.

The LCC's pre-war estates had been heavily criticised but, given the requirement to accommodate 136 people per acre, the solution initially appeared to lie in the same unsatisfactory model. Tufnell Park Estate (1946) was built with stairs rather than lifts, and with old-fashioned domestic equipment and gallery access "which eroded privacy in the kitchens, bathrooms, and bedrooms which looked on to it".[43] To popularise flats, Forshaw introduced eight-storey blocks with lifts at Woodberry Down, Hackney (1946–48), but, since this slightly raised the cost and reduced the number of dwellings, it was only allowed on the understanding that it would not set a precedent.[44] It was soon to be followed by the Council's two most prestigious estates in London, Ackroydon in Wimbledon (1950–53) and Alton East Estate, Roehampton (1951–54), both mixed developments with point blocks (adapted from Swedish initiatives), and by others which closely followed the model of Le Corbusier, as well as by various types of housing set in parkland with mature trees. Other notably successful public housing included Powell and Moya's multi-phase mixed development at Churchill Gardens (1946–50): ". . . the Pimlico blocks . . . provide a landmark vying with any the twentieth century provided in London".[45] At Spa Green Estate, Rosebery Avenue (1946–50; fig. 19), criticised for overly decorative tiling – recalling criticism of Lubetkin's pre-war Highgate flats – and at Priory Green Estate, Finsbury (1952), Tecton drew on pre-war Continental models to develop a thoroughly modern "architectural language for the block of flats".[46] Gibberd, much of whose work on large developments was carried out in East London, also designed the first phase (1951) of the Regent's Park Estate, work extended in 1952–53 by Davies & Arnold with a combination of housing and eleven-storey blocks of flats (fig. 20). This was the territory for Auerbach's *Building Site, St Pancras* (cat. 4).

Lionel Esher was to summarise the architectural character of early post-war housing as "hard" or "soft": Churchill Gardens and Alton West Estate (1955–59; fig. 21), which reflected the impact of the Marseilles Unité, were "hard", whereas "cottagey Lansbury" and mixed developments like Ackroydon, which incorporated old landscaping, represented the "soft" option.[47] But despite very disparate architectural approaches, after the first panic-stricken response to the housing crisis, developments were designed above all to improve people's living conditions. Priory Green was exemplary but not unique in incorporating a laundry, nursery

Fig. 20
Regent's Park Estate, with the nineteen-storey Armstrong & MacManus block in the foreground and the eleven-storey Davies & Arnold block in the centre; the complete estate developed 1951–63
Photograph: English Heritage, National Monument Record

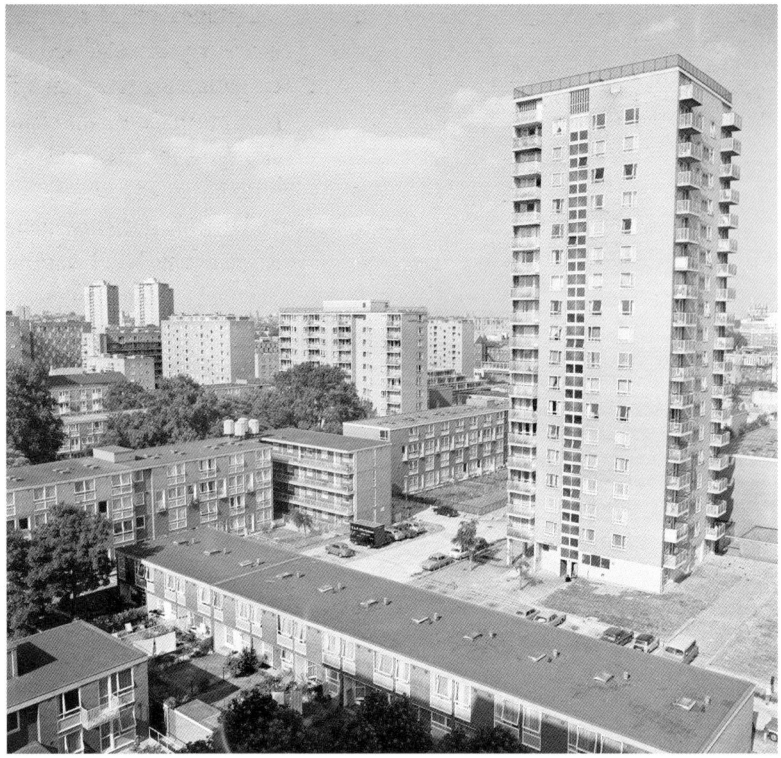

school, community centre and pub in response to the tenants' expectations of improved facilities.[48] In the mid 1950s the LCC was to extend the concept of facilities to include works of art, through a very successful scheme to commission sculpture and murals for its properties, which extended from primary schools to old-age homes.[49]

Virtually all the LCC's schools had been damaged or destroyed and many new buildings were needed to deliver the child-centred education envisaged by the 1944 Education Act. A widely adopted model involving prefabricated steel frames and a modular design system was developed by Hertfordshire County Council to build schools that were varied in scale and plan and judged "important as architecture".[50] The LCC adapted the Hertfordshire system for primary schools but allowed architects more individual decisions in the design of "the more prestigious ... comprehensive", resulting in buildings of varied quality.[51]

Despite a desperate need for an industrial revival, the "second task of the reconstruction program ... new buildings for manufacturing industries" was, as Barnett has argued, the Cinderella of the reconstruction.[52] It took place virtually without architects and is thus almost unrecorded. Though almost no new factories were erected for a year after the war, they were increasing fourfold by September 1948 as regulation only applied to buildings over 5,000 sq. ft, leading one writer to comment on the likelihood of a "steady progress ... at the rate of 4,999 sq. ft a year".

Fig. 21
Alton West Estate, Roehampton
(LCC Architects' Department), 1954–58
Photograph: Conway Library,
The Courtauld Institute of Art

Existing factories, many of which had simply been rebuilt by taking advantage of a legal loop-hole, were entirely free from controls.[53] A rare exception to invisibility was the Brynmawr rubber factory in South Wales (Architects Co-Partnership, 1947–51). Designed to be architecturally outstanding, it was considered sensational, its main production area roofed with flattened concrete domes, "the largest of their kind in the world".[54]

Of the few distinguished public buildings erected in these years, the Royal Festival Hall was the first, designed by Lesley Martin for the LCC. Opened for the inauguration of the Festival of Britain on 3 May 1951, it has been described as "the most successful public building in post-war Britain", an "aloofly monumental" example of the excellence of the LCC Architects' Department.[55] The second great public building was the Trades Union Congress Memorial Building in Great Russell Street, by David du R. Aberdeen (1956–57). Both an elegant, sophisticated conference building with no concessions to outmoded 'styles' and a receptacle for Epstein's great memorial sculpture, it was spectacularly successful: the Festival Hall was "the only comparable building erected in London since the war".[56]

The grandest and most sensitive of all public buildings remained St Paul's. The Barlow and Uthwatt reports had conferred extensive powers on local authorities, including the Corporation of London, to buy land

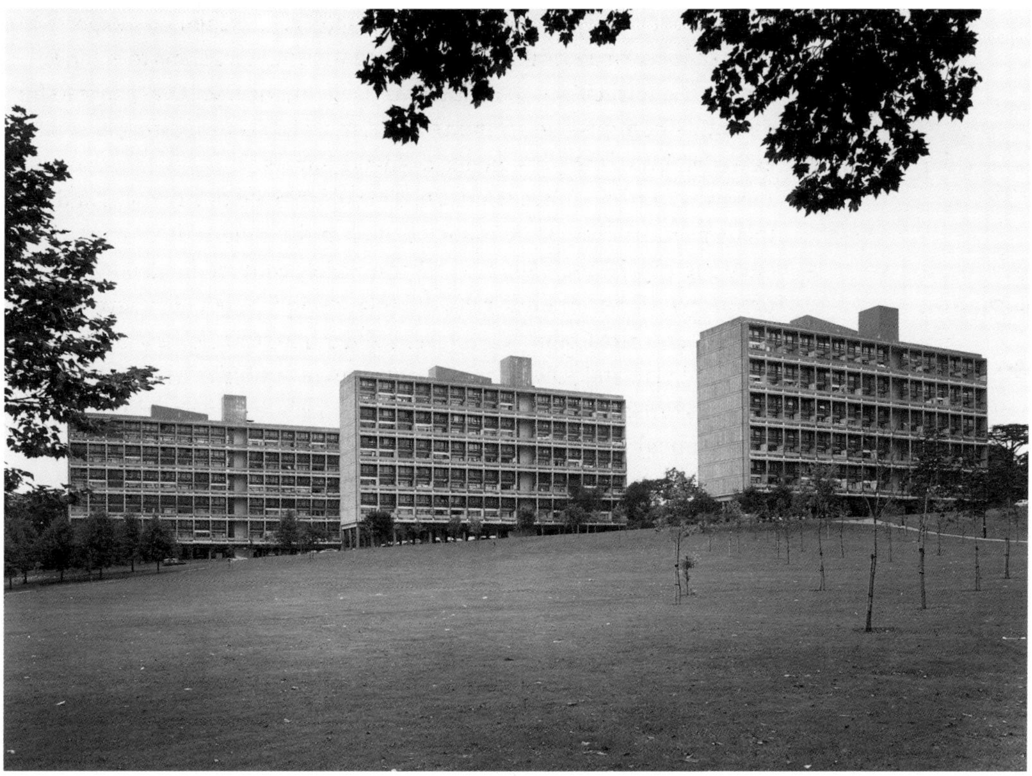

and amend previous developments. The Corporation applied for all the land around St Paul's except the south-west quadrant. In their final report as planning consultants, C.H. Holden and W.G. Holford wrote commandingly that the view of St Paul's revealed by bomb damage "should never be wholly lost in the rebuilding".[57] Much editorial space was devoted to the architectural possibilities of the most symbolically important site in London but the buildings that came to dominate the area to the east, which was to be the locus for Auerbach's *Building Site near St Paul's* (cat. 6), both had the blandest of neo-Georgian façades – Gateway House (Trehearne & Norman, Preston & Partners, 1953) and New Change House (Victor Heal & Smith, 1953–60; fig. 22), the latter 'lessor scheme' offices for the Bank of England.[58] In 1955 Sir William Holford was appointed to formulate a plan "to improve and increase the variety of viewpoints [of the Cathedral] for those who move about the precinct… and to achieve a sense of enclosure… by creating a whole series of smaller and related open spaces".[59] Richards pleaded for robust, imaginative design in relation to the Cathedral, but Holford's Paternoster Square, elegant in drawings, was bleak and empty in practice.[60] Remodelled in the 1990s with busy post-modern detail, it is heavily used as a public space if not greatly admired.

Deregulation – Diversity – Goldfish Bowls

The end of building controls gave free rein to developers, whose approach to architecture, primarily determined by construction systems, led Esher to distinguish "industrialized building from 'creative' designers".[61] As early as 1954 Richards referred gloomily to "piecemeal" building schemes and the possibility of "complete failure" in the reconstruction effort, citing the sterility of neo-Georgian and neo-Renaissance, default styles in a design approach where architects were "completely in the stranglehold of big business".[62] As an alternative he commended Bucklersbury House (Campbell-Jones & Son, 1954), which was distinguished by set-backs rather than interior courts or light wells, as among "the most promising of the post-war City projects" and an unquestionably modern building.[63] "Thereafter," Esher comments, "stubby skyscrapers became the norm."[64]

Early in 1956 the *Architectural Review* enthusiastically proclaimed: "Architects are engaged on a greater variety of buildings than at any time for nearly twenty years".[65] Churches, airports – including Heathrow – laboratories, an atomic power station, colleges and department stores were in progress across the country, indicating a swathe of new patrons. These were predominantly property investment companies, whose value as patrons was summarily dismissed and their pursuit of profitable office-space compared with the earlier lessor scheme. Rose, himself a developer, described their design and construction approach: "… the

Fig. 22
New Change House, St Paul's (Victor Heal & Smith), 1953–60
Photograph: Conway Library, The Courtauld Institute of Art

architects who were most popular were those whose designs and specifications tended to provide the maximum amount of lettable floor-space, the essence of attainable maximum rent: sacrificing elaborate façades in favour of curtain wall construction and limiting floor-to-ceiling heights to the minimum permitted by building byelaws".[66] Rose acknowledged certain exceptions, namely Marylebone House and Woolworth House, "an excellent example of mid 1950s stripped classicism", with Castrol House (Melvin Ward & Partners with Casson & Conder, 1955–60) "the earliest slab-and-podium tower with curtain walling in Britain, echoing Lever House [in New York]."[67] Like many corporate buildings in the 1960s, Castrol House contained a work of art in the entrance hall, in the form of a monumental aluminium relief by Geoffrey Clarke.[68] Nevertheless Rose concluded: "When I look at the streets of London and at the buildings that were put up in that era, I'd say we made a muck of it".[69]

A period of heavy demand for office space followed in the early 1960s. Two corporate headquarters, the Vickers Tower (Ronald Laird & Partners, 1963, fig. 25) and the Shell Centre (Sir Howard Robertson and Maynard

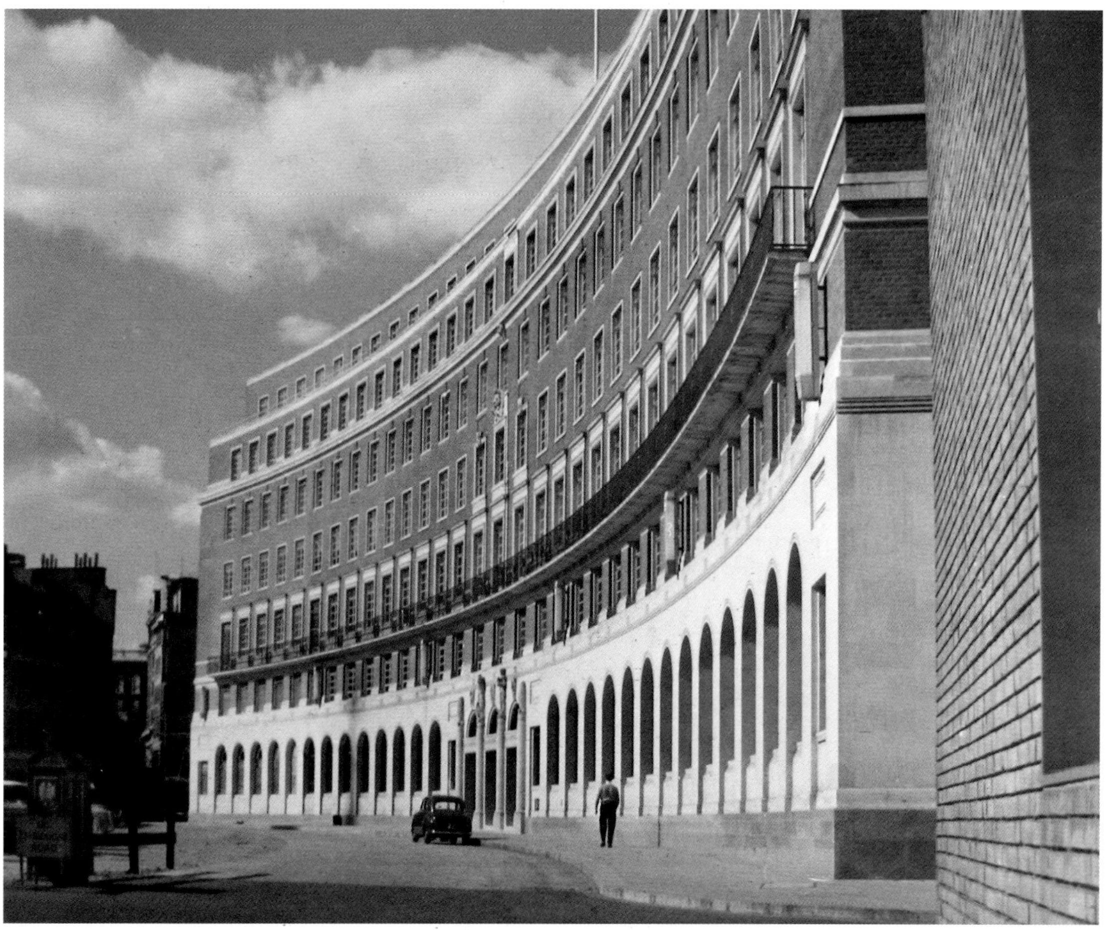

Fig. 23
Shell Building, South Bank, under construction, *c.* 1961
Photograph: Sir Robert McAlpine Ltd

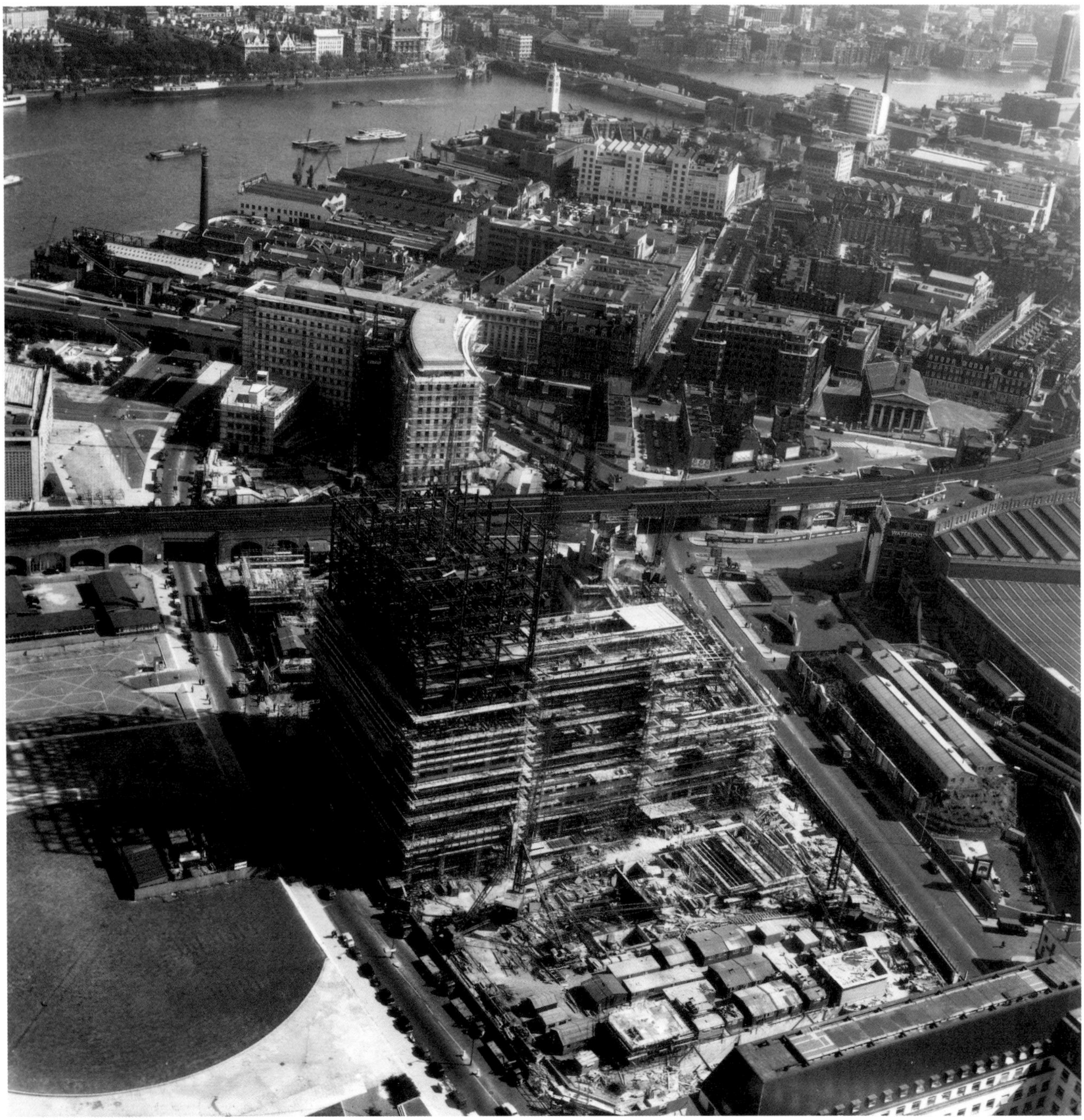

Smith for Easton, Robertson, Cusdin, Preston & Smith, 1953–63; figs. 23 and 24), on opposite banks of the Thames, appear to represent different eras, the Tower gleaming and dynamic while the Shell Centre has been described as "the saddest failure on the South Bank".[70] A self-effacing complex on part of the former South Bank exhibition site, it was the origin of Auerbach's three *Shell Building Site* paintings (cat. 7–9). Government offices, including the Board of Trade and its successors, have occupied No. 1 Victoria Street (Ronald Fielding, 1959–64) since its completion. The locus of *Victoria Street Building Site* (cat. 10), it was refurbished by the design consultancy DEGW in 1994–95, though the addition of a fashionable glass foyer has done little to alleviate the monumental blankness of the street frontage.

The reconstruction of the West End was more varied and revealed a higher design standard than the City, though the lack of a master plan led Ian McCallum to describe the area as "the *victim* of a building boom".[71] Signs of prosperity emerged surprisingly early with the already mentioned Time–Life Building (1951–53), the locus of *Building Site, Bruton Street* (cat. 2). The first post-war office building designed for prestige on a prime West End site, it was a showcase for contemporary art, with Henry Moore's *Screen* on the Bond Street façade. Ceramics by Lucie Rie and a mural by Ben Nicholson were among the works of art in the interior, while a small army of consulting architects and designers contributed to the opulent furnishings. No City office approached this degree of luxury, not least because high-status offices and corporate headquarters in the 1950s and 1960s tended to be designed on the model of the "prestige pancake" – low-rise, extensive, and thus unsuitable for the inner city.[72]

Oxford Street Building Site I and *II* (cat. 11a and 11b) refers to the rebuilding of the John Lewis department store (Slater & Uren, 1955–60), established in Oxford Street in 1895 but irreparably damaged in 1940. The new building, elegant and resolutely modern, remains one of the most successful large shops of the period, with Barbara Hepworth's *Winged Figure* on its eastern façade. The vast, venerable Maple's Furniture Store on Tottenham Court Road, subject of *Maples Demolition* (cat. 12), was entirely rebuilt by Dalgleish & Pullen (1950–59), but after some difficult years went into receivership in 1997. The building has since been heavily remodelled. The growth in luxury shops in the West End was acknowledged by the *Architectural Review*'s publication in 1957 of a lavishly illustrated article on shop display, to be followed regularly by reports on showroom interiors.[73] When the new Sandersons Wallpaper and Fabric showroom (Slater & Wren with Geoffrey Holroyd) opened in Berners Street in 1960 it was a prominent symbol of the new prosperity, with a marble and glass mosaic in the entrance hall, a sculpture by Jupp Dernbach Mayers in the courtyard and a stained-glass window by John Piper and Patrick Reyntiens on the staircase.[74] The last of Auerbach's building-site paintings, *Rebuilding the Empire Cinema, Leicester Square* (cat. 13) refers to a building that opened in 1884 as a live theatre, then

Fig. 24
Shell Building, South Bank (Sir Howard Robertson & Maynard Smith), 1953–63
Photograph: Sir Robert McAlpine Ltd

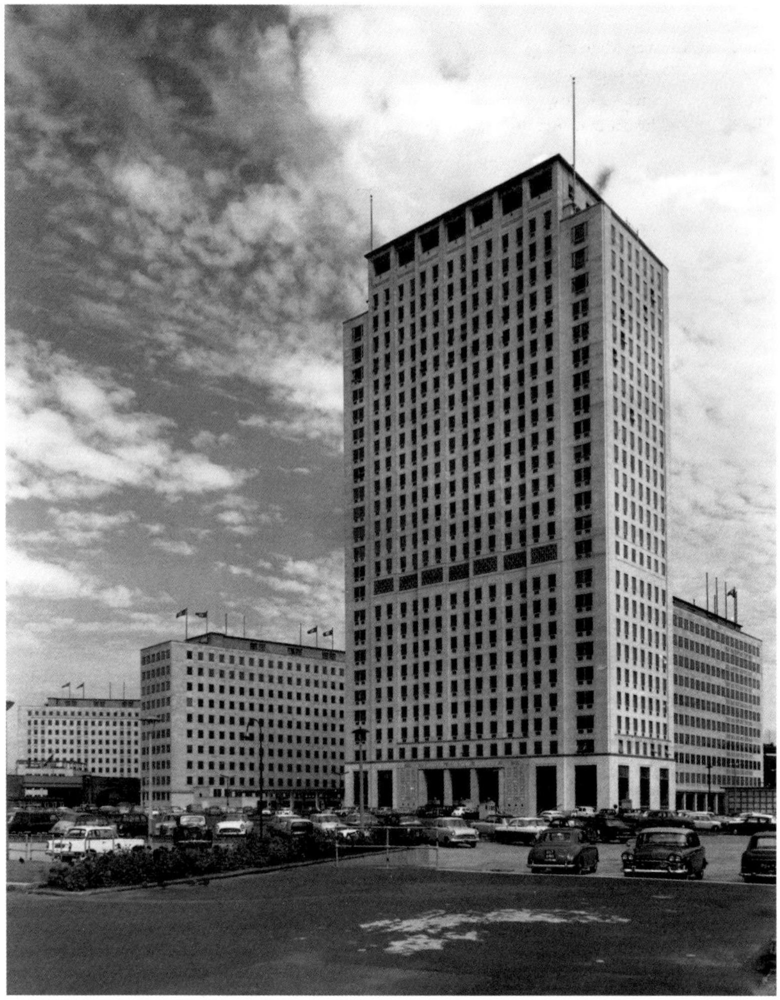

was totally redesigned by George Coles in 1962, with a 63-ft-long Perlux screen and 'lounger chairs'.

Summarising a decade of post-war building in western Europe in 1957, Richards, pessimistic as ever, deplored the "meanly finished utility commercial style" of much British building, the "abysmally low standard" of office building and, ominously, "the return of the speculative house-builder".[75] Yet only two years later he found that traditional firms had successfully embraced "modern architecture", with the result that it had become a style, albeit one with "an almost bewildering variety of form and structure".[76] Thus Richards acknowledged the inevitability of the dilution of London's reconstruction plans. Totalising plans for its future had been drawn up at a time when their optimism had no doubt provided comfort, but, once the dispossessed were rehoused and the economy was recovering, complete renewal was neither practical nor compelling.

A point constantly reiterated by the *Architectural Review* but difficult to convey to an architecturally conservative and largely indifferent public

Fig. 25
The Vickers Tower, Millbank
(Ronald Laird & Partners), 1963
Photograph: Conway Library,
The Courtauld Institute of Art

was that style carried meaning. Despite the efforts of architectural critics, neo-Georgian remained for a considerable time (and it is resurgent today) the only style widely acknowledged as culturally significant, though prefabricated frames and glass were eventually recognised as emblems of the technological imperative, efficiency and productivity. The record of London's reconstruction since 1945 reveals it as both a monument to high capitalism and the site of millions of small homes. It is contradictory and discordant and much of its architecture is mundane but its hybridity is eloquent. As Peter Ackroyd has written, "London has always been an ugly city ... it has always been rebuilt, and demolished, and vandalised"; moreover, it has not much welcomed *grands projets*.[77] The result is a 'piecemeal' city with a visual diversity and complexity that accord – fortuitously – with the *Architectural Review*'s long-ago calls for architects to reassert a picturesque sensibility.

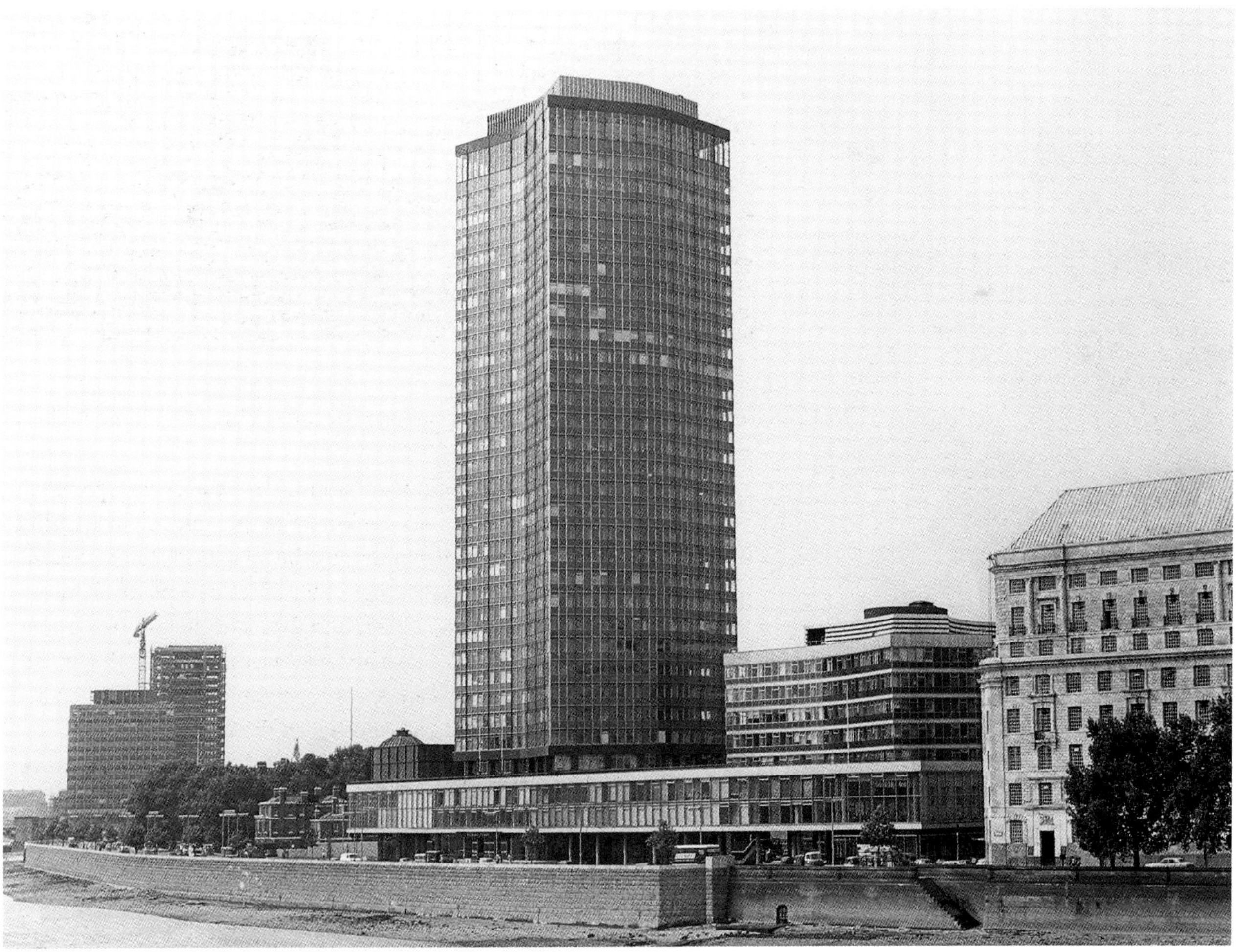

Notes

1 Doris Lessing, *The Four-Gated City*, London, 1969, pp. 19–20.
2 They are perhaps best characterised by a modern understanding of 'Gothic', as a genre that addresses "fear, anxiety and death": Gilda Williams, 'Gothic v Gothick', *Art Monthly*, no. 326, May 2009, pp. 7–10.
3 Richard Sheppard, *Building for the People*, London (George Allen & Unwin Ltd) 1948, p. 16.
4 J.H. Forshaw and Patrick Abercrombie, *City of London Plan, Prepared for the London County Council*, London, 1943, p. 99.
5 J.M. Richards, 'Rebuilding the City', *Architectural Review*, vol. 115, no. 690, June 1954, pp. 378–85.
6 New Ideal Homesteads, Laing & Costain, Taylor Woodrow, Wates and Wimpey were the most prominent; G.E. Cherry, *Cities and Plans: the Shaping of Urban Britain in the Nineteenth and Twentieth Centuries*, London (Edward Arnold) [1988], 1993, p. 94.
7 *Ibid.*, pp. 88-89.
8 It excluded war damage, which was carried out by labour employed through local authorities and utility companies on a 'pool' system, transferable between building firms; see Ian Bowen, 'The control of building', in B.N. Chester (ed.), *Lessons of the British War Economy*, Cambridge, 1951, pp. 122–43.
9 The Modern Architectural Research Group, chaired by Arthur Korn; see Cherry 1988, p. 112.
10 Editorial, 'Exterior furnishing or Sharawaggi: the art of making urban landscape', *Architectural Review*, vol. 95, no. 565, January 1944, pp. 3–8; Anthony Jackson, *The Politics of Architecture. A History of Modern Architecture in Britain*, London (Architectural Press) 1970, pp. 161, 162.
11 Cherry 1993, p. 116.
12 Jackson 1970, p. 162. "[Abercrombie's proposals] may perhaps be reckoned a fairly shrewd anticipation of the measures democracy may be ready to adopt when the war is over" (comment on Abercrombie from John Dower, 'The Barlow Report'. *RIBA Journal*, no. 47, p. 101; quoted in Jackson 1970, p. 80).
13 Forshaw and Abercrombie 1943, p. 3.
14 *Ibid.*, p. 79.
15 *Ibid.*, p. 83. The standard densities were 100, 136 and 200 people per acre.
16 Jackson 1970, pp. 166, 169.
17 Correlli Barnett, *The Audit of War: The Illusion and Reality of Britain as a Great Nation*, [1986], London (Macmillan) 1987, p. 8.
18 *Ibid.*, p. 242.
19 *Ibid.*, p. 243.
20 'Address by the Minister of Works', *RIBA Journal*, no. 52, 1945, p. 215, quoted in Jackson 1970, p. 168.
21 Memorandum, CAB 87/10, R(45)8, cited in Barnett 1987, p. 246.
22 Nicholas Bullock, *Building the Post-war World: Modern Architecture and Reconstruction in Britain*, London and New York (Routledge) 2002, pp. 191, 153.
23 Sheppard 1948, p. 20.
24 *Ibid.*, p. 19.
25 *Ibid.*, pp. 27, 20.
26 Justin Davis Smith, *The Atlee and Churchill Administrations and Industrial Unrest, 1945–1955. A Study in Consensus*, London & New York, 1990, p. 4.
27 J.W. Duncan, W.E.J. McCarthy, G.P. Redman, *Strikes in Post-war Britain: A Study of Stoppages of Work due to Industrial Disputes, 1946-73*, London, 1983, pp. 35, 61, 96.
28 Jack Rose, *The Dynamics of Urban Property Development*, London and New York (E & F.N. Spon) 1985, pp. 149, 150.
29 Shirley Green, *Who Owns London?*, London, 1986, p. 97.
30 Rose quoted in S. Humphries and J. Taylor, *The Making of Modern London 1945–1985*, London, 1986, p. 61; Rose 1985, p. 160.
31 Rose 1985, p. 160.
32 *Ibid.*, p. 155.
33 Editorial, *Architectural Review*, January 1944; I. de Wolfe, 'Townscape: A Plea for an English Philosophy Founded on the Fine Rock of Sir Uvedale Price', *Architectural Review*, vol. 106, no. 636, December 1949, pp. 355–62; Bullock 2002, p. 84.
34 'Foreword', *Architectural Review*, vol. 110, no. 656, August 1951, Special Issue on the South Bank Exhibition, pp. 73–79; for Somerford, see Bullock 2002, p. 84.
35 'New Empiricism: Sweden's latest style', *Architectural Review*, vol. 101, no. 606, June 1947, pp. 199–203; Bullock 2002, pp. 45–47; see also the explanation in Reyner Banham, 'The New Brutalism', *Architectural Review*, vol. 118, no. 708, December 1955, pp. 355–58, 361.
36 Jackson 1970, p. 181.
37 Such as Trehearne & Norman Preston & Partners and TP Bennett & Son: Lionel Esher, *'A Broken Wave': The Rebuilding of England 1940–1980*, London (Pelican) 1981, p. 59; Bullock, 2002, p. 256.
38 Jackson 1970, p. 182; Banham 1955. In 1953 Banham joined the *Architectural Review* and Theo Crosby joined *Architectural Design*; they both supported the Smithsons.
39 David Kynaston, *Austerity Britain 1945–51*, London, (2007) 2008, pp. 19–59.
40 See the intermittent series in *Ideal Home* 1941–45, 'Planning for the Future' and a lecture series, Royal Society of Arts, November 1941–May 1942, published as *The Post-War Home: Its Interiors and Equipment*; cited in David Jeremiah, *Architecture and Design for the Family in Britain, 1900–70*, Manchester and New York, 2000, pp. 125–26.
41 Esher 1981, p. 48.
42 Aileen Tatton-Brown, 'County of London Plan', *Architectural Review*, vol. 105, no. 629, May 1949, pp. 223–30.
43 Bullock, 2002, p. 207.
44 *Ibid.*, p. 211. Woodberry Down has been described as "the most interesting

development built by the LCC between 1945 and 1950": *ibid.*

45 Henry Russell Hitchcock, 'Criticism: Pimlico', *Architectural Review*, vol. 114, no. 681, September 1953, pp. 176–84.

46 Bullock 2002, pp. 87, 88.

47 Esher 1981, p. 107.

48 'Flats at Finsbury', *Architectural Review*, vol. 112, no. 670, October 1952, pp. 241–52.

49 Margaret Garlake, '"A war of taste": the LCC as art patron 1948–1965', *The London Journal*, vol. 18, no. 1, 1993, pp. 45–65.

50 Richard Lewelyn Davies and John R. Weeks, 'The Hertfordshire Achievement', *Architectural Review*, vol. 111, no. 666, June 1952, pp. 367–87; Bullock 2002, pp. 183, 185–86; Andrew Saint, *Towards a Social Architecture: The Role of School Building in Post-war England*, New Haven and London (Yale) 1987, pp. 58–111.

51 Saint 1987, p. 186.

52 Jackson 1970, p. 166.

53 Tatton-Brown 1949.

54 R. Furneaux Jordan, 'Brynmawr factory in South Wales', *Architectural Review*, vol. 111, no. 663, March 1952, pp. 143–64. The factory was demolished in 2002 after protracted efforts to save it.

55 Peter Carolin and Trevor Dannatt (eds.), *Architecture, Education and Research. The Work of Lesley Martin: Papers and Selected Articles*, London, 1996, p. 33; Jackson 1970, p. 178.

56 'T. U. C. Memorial Building', *Architectural Review*, vol. 122, no. 731, December 1957, pp. 370–87.

57 C.H. Holden and W.G. Holford, *The City of London: A Record of Destruction and Survival* (Final Report of the Planning Consultants), 1947, London, 1951, p. 198.

58 The New Change site is currently (2009) being redeveloped by Land Securities with an immense curtain walled structure as the headquarters of K & L Gates LLP.

59 'St Paul's', *Architectural Review*, vol. 119, no. 713, June 1956, p. 297.

60 J.M. Richards, 'City Rebuilding: 2', *Architectural Review*, vol. 116, no. 693, September 1954, pp. 145–47; 'St Paul's' (as note 59), 1956, pp. 295–98.

61 The most prominent developers included Sir Charles Clore, Jack Cotton, Lew Hammerson, Joe Levy, Lord Samuel. Samuel "pioneered the full repairing and insuring lease, whereby tenants undertook responsibility for all outgoings; and Joe Levy pioneered shorter leases with regular rent reviews ... [and] ensured the profitability of commercial lettings through times of rising repair costs and inflation, in sharp contrast to the fate of residential lettings". Lord Samuel was Chair of Land Securities, the largest publicly listed property development company in the UK in 1985; see Green 1986 pp. 97, 98; Esher 1981, p. 66.

62 J.M. Richards, 'Rebuilding the City', *Architectural Review*, vol. 115, no. 690, June 1954, pp. 378–85.

63 *Ibid.*, p. 383. The Royal Fine Art Commission had "objected to its bulk, but was overruled by the Minister of Housing and Local Government, and the client showed preferences for an academic style of design with neo-classical embellishments"; *Architectural Review*, vol. 115, no. 685, special preview issue, January 1954, pp. 71–72.

64 Esher 1981, p. 114.

65 'Foreword', *Architectural Review*, vol. 119, no. 708, January 1956, pp. 5–7.

66 Rose 1985, p. 153.

67 Cited in Humphries and Taylor 1986, p. 62; details in '20th century landmarks on Marylebone Road protected', *C20*, Spring 2009, p. 15. All three buildings have recently been incorporated into the Dorset Square Conservation Area.

68 Geoffrey Clarke, *The Extraction and Refining of Oil*, 1959. For industrial and commercial patrons, see Alan Osborne (ed.), *Patron. Industry Supports the Arts*, London, 1966.

69 Quoted in Humphries and Taylor 1986, p. 62.

70 Esher 1981, p. 110.

71 Ian McCallum, 'West End offices', *Architectural Review*, vol. 120, no. 717, October 1956, pp. 219–24.

72 Elain Harwood, '"Prestige pancakes": The influence of American planning in British industry since the war', *Industrial Architecture*, London, 1994, pp. 77–89.

73 Herbert Tayler, 'Shops', *Architectural Review*, vol. 121, no. 721, February 1957, pp. 99–111.

74 *Architectural Review*, vol. 125, no. 744, special preview issue, January 1959, pp. 121–27 on Thorne House and pp. 405–09 and 420–21 on Sandersons.

75 J.M. Richards, 'Europe Rebuilt: 1946–56', *Architectural Review*, vol. 124, no. 722, March 1957, pp. 158–76.

76 'Foreword', *Architectural Review*, vol. 123, no. 732, January 1958, pp. 5–8.

77 Peter Ackroyd, *London: The Biography*, London (Chatto & Windus) 2000, p. 760. On *grands projets* Francis Sheppard cites only the Barbican and the Docklands regeneration in recent years; see *London: A History*, Oxford, 1998, p. 351; see also Richard Rogers and Mark Fisher, *A New London*, London, 1992.

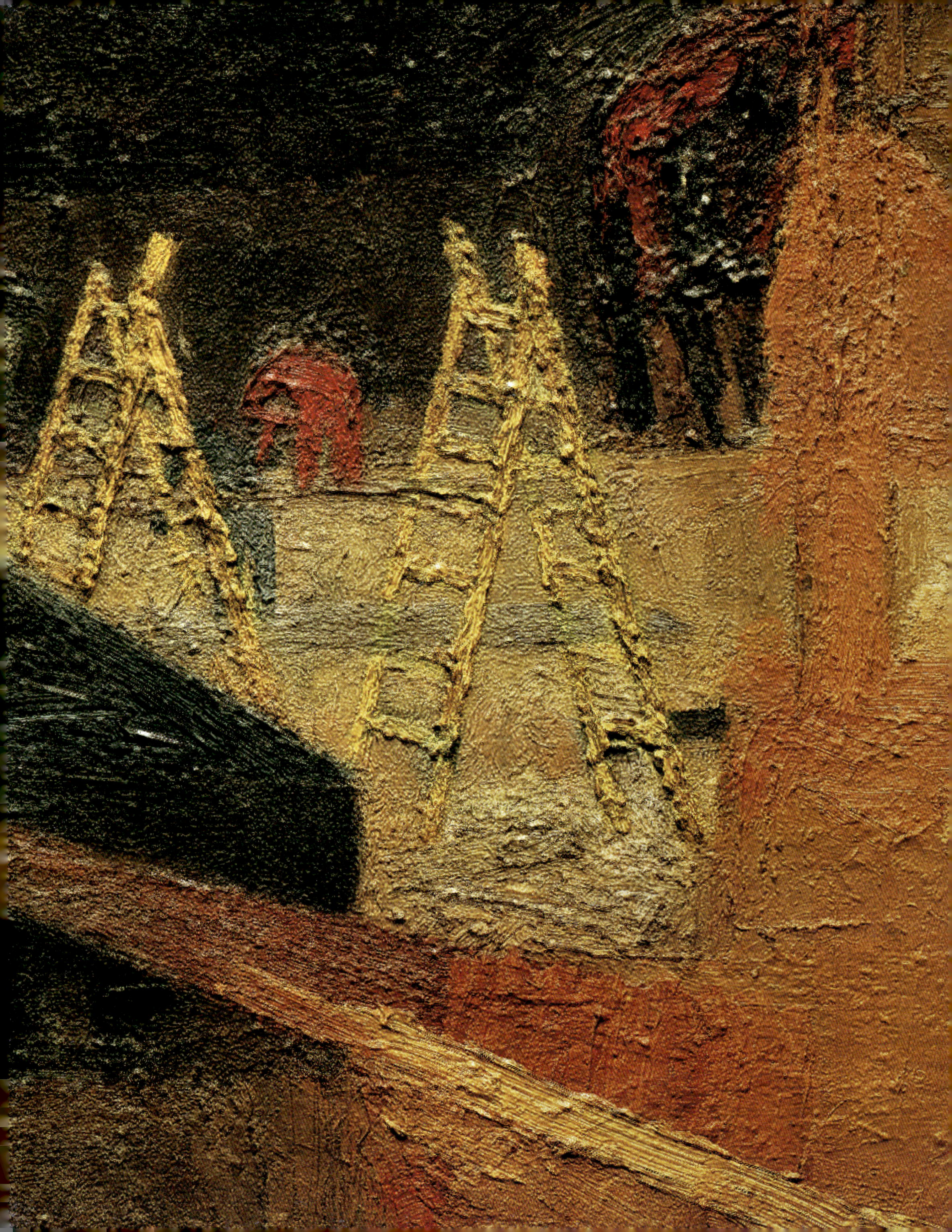

A Human Universe: Auerbach's Building-Site Paintings and Existentialism

PAUL MOORHOUSE

For almost sixty years a surprisingly restricted range of places in London has nourished the development of Frank Auerbach's art. His portraits and paintings of the figure occupy a numerically larger proportion of his relatively modest output and are its vital core. But, alongside this preoccupation with depicting individuals, his parallel engagement with making drawings and paintings of particular London locations has provided an essential wider context for exploring issues connected, no less fundamentally, with human existence (fig. 26).

From the outset, immersing himself in the early 1950s in the ruined post-war landscape of London, Auerbach recognised his immediate surroundings as "the fabric of one's life".[1] The areas of bombed land then being slowly and doggedly rebuilt as the capital sought to reassert itself yielded a ready supply of subjects close to hand. In that sense, the building-site paintings that Auerbach made between 1952 and 1962 are a direct response to the world he occupied. But, in parallel with his direct contact in the studio with a human presence (fig. 27), his early involvement with images of dereliction and renewal may be seen as arising from his recognition of the potential of such subjects as armatures that sustain a radical visual transformation. This process of transformation, in which observation and a subjective, internalised response to the world were enmeshed, transcended the imperative simply to depict.

Auerbach's approach had its roots in the intellectual and artistic fabric that animated London during the 1950s. The period was dominated by existentialist ideas, imported from Paris, which, through discussion, debate and subsequent appropriation by the Press, acquired an ever-widening currency. At the heart of this philosophical outlook was the premise that

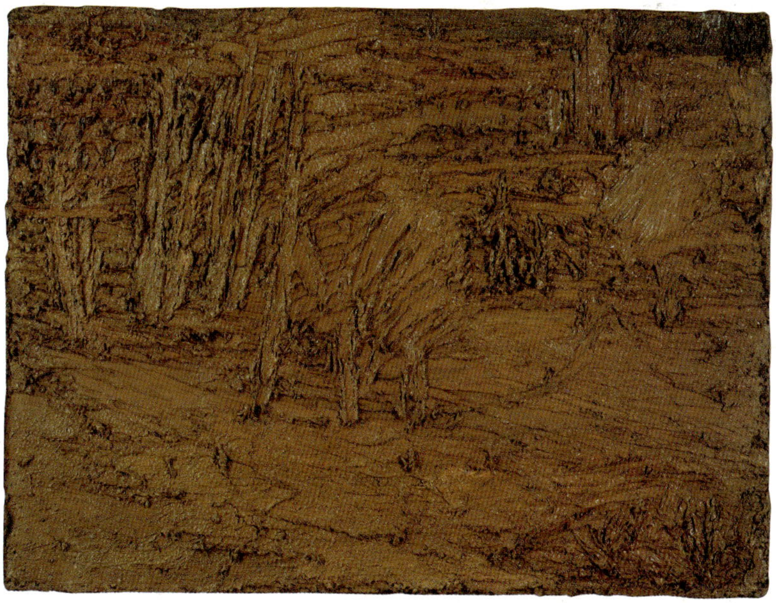

Fig. 26
Frank Auerbach, *Primrose Hill*
1954–55
Oil on board, 86.4 × 116.8 cm
Private collection

OPPOSITE
Detail cat. 1

Fig. 27
Frank Auerbach, *E.O.W. Nude*
1953–54
Oil on canvas, 62.2 × 76.8 cm
Private collection

an individual's experiences and actions not only formed the essential components of life: at a fundamental level this active engagement with the world provided meaning, understanding and purpose in a godless universe.

Auerbach's building-site paintings constitute a distinct body of work. In their progressive, radical reworking of the observed motif they convey a keen sense of its independent character – its otherness. As such, they mark the beginning of Auerbach's life as a painter. But the profound subjectivity of his involvement with themes of excavation and reconstruction was not only visual: it was an inescapable aspect of the way the images came into being and were resolved. In this respect the paintings may also be understood in relation to existentialist ideas. Viewed in that light, they can be regarded as a kind of crucible in which the essence of his art was forged.

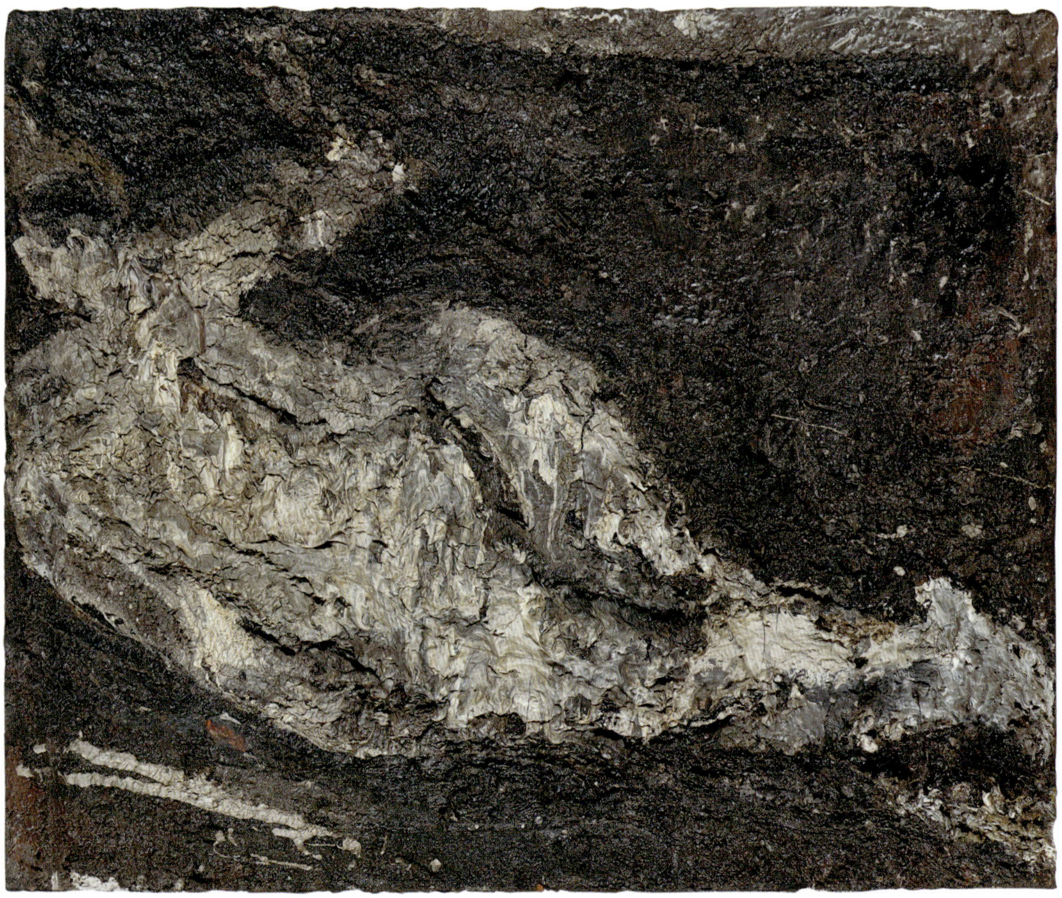

Fig. 28
Frank Auerbach as a student at the LCC Open-Air Exhibition, Embankment Gardens, London, July 1948
Photograph by Daniel Larson

A Marvellous Landscape

As an eight-year-old refugee from Nazi persecution in his native Germany, when Auerbach arrived in England in April 1939 his first impressions of his new home in Kent were that he had been "picked up and transported into a different world".[2] When he came to live in London in 1947 this sense of the strangeness of his surroundings must have deepened. Auerbach found a city bearing the scars of a costly and destructive conflict. Having been accepted at St Martin's School of Art he commenced his studies in September 1948 (fig. 28). Often walking to his destination, he encountered bomb sites, shattered buildings and expanses of excavated ground. Both *en route* and further afield, the devastation was everywhere apparent, the centre of the capital, the City, the East End and areas close to the River being particularly badly affected. A startling fabric of disorder was the backdrop to his daily life – the wreckage of architecture and geological chaos being clearly visible from the street, so that it permeated the everyday, acquiring a curious familiarity. But Auerbach seems to have viewed his surroundings with eyes undimmed by nostalgia for lost landmarks. Where surviving inhabitants saw wreckage, Auerbach perceived "a marvellous landscape"[3] of mountains, precipices and crags – a formal drama imposing in its austerity and rich visual incident. For the young artist, London's landscape, pulverised by bombs, provided an irresistible subject.

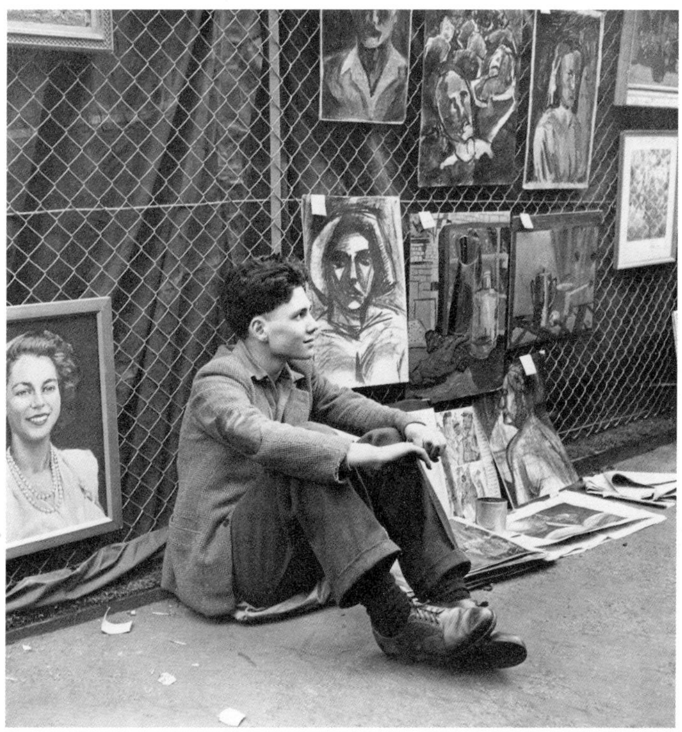

The programme of rebuilding and renewal that gradually got underway found an intellectual corollary in London's art schools in the early 1950s. At St Martin's Auerbach encountered a lively, if rather serious, atmosphere generated by students wanting to create art for their own time. In addition to the usual new student intake, several of Auerbach's contemporaries were ex-servicemen, slightly older than himself, whose attitudes were inevitably coloured by their war-time experiences. For younger students also, the unfolding post-war situation, with its revelations of the Holocaust and the advent of atomic warfare, generated a sense that old values no longer applied. For many, the styles and ideas of the previous generation of artists, which had been dominated by neo-Romanticism, seemed out of date and out of touch with the world as it now was. The new atmosphere was that art had to be grounded in tangible experience rather than in idealised pastoral fantasy. For students of Auerbach's generation, among them Jack Smith, who attended St Martin's in 1948–50, this meant an art that responded to ordinary, everyday reality. Smith observed, 'The art of the 1940s in France and England didn't excite me. I reacted against its lack of creative and inventive energy. So, like many artists before me I turned to my own environment for subject matter.'[4]

In 1949, during Auerbach's second year at St Martin's, he met Leon Kossoff, who had just started at the school, having recently been demobbed. Although Kossoff was older than Auerbach by five years, the two struck up a close friendship. Drawing from life was regarded by both as an important and essential activity. Sharing a mutual regard for each other's efforts, Auerbach and Kossoff found a reciprocal willingness to assist one another's wish to work from the model. From the early 1950s it became their custom to spend a day a week during which one would sit while the other drew and painted. Then they would swap over, alternating throughout the day.

Two surviving portraits of Auerbach completed by Kossoff in 1953 – one a head study, the other a half-length – demonstrate the rich earthy palette and extreme textures that are the principal characteristics of his work at this time. In Auerbach's drawings and paintings of Kossoff, which he continued to make over the next four years, there is similar physical evidence of images which have evolved over a long period of time, having been continually reworked. His *Head of Leon Kossoff* (1954) is thickly painted but its rutted surface is surprisingly animated, as if bearing the traces and tracks of endless, energetic movement (fig. 29). A related drawing of Kossoff, dated 1957, suggests the negative of the painting, its surface abraded and scarred by constant revision and erasure. In their different ways, both these images by Auerbach yield a sense of a complex, accumulated and urgent involvement in giving physical substance to the fleeting and ephemeral.

The way that such images seem not only traversed by activity, but also to have been wrested from their own fabric, has a parallel in the drawing expeditions to outdoor sites which Auerbach and Kossoff shared around

Fig. 29
Frank Auerbach, *Head of Leon Kossoff*
1954
Oil on board, 41.3 × 31.7 cm
Private collection

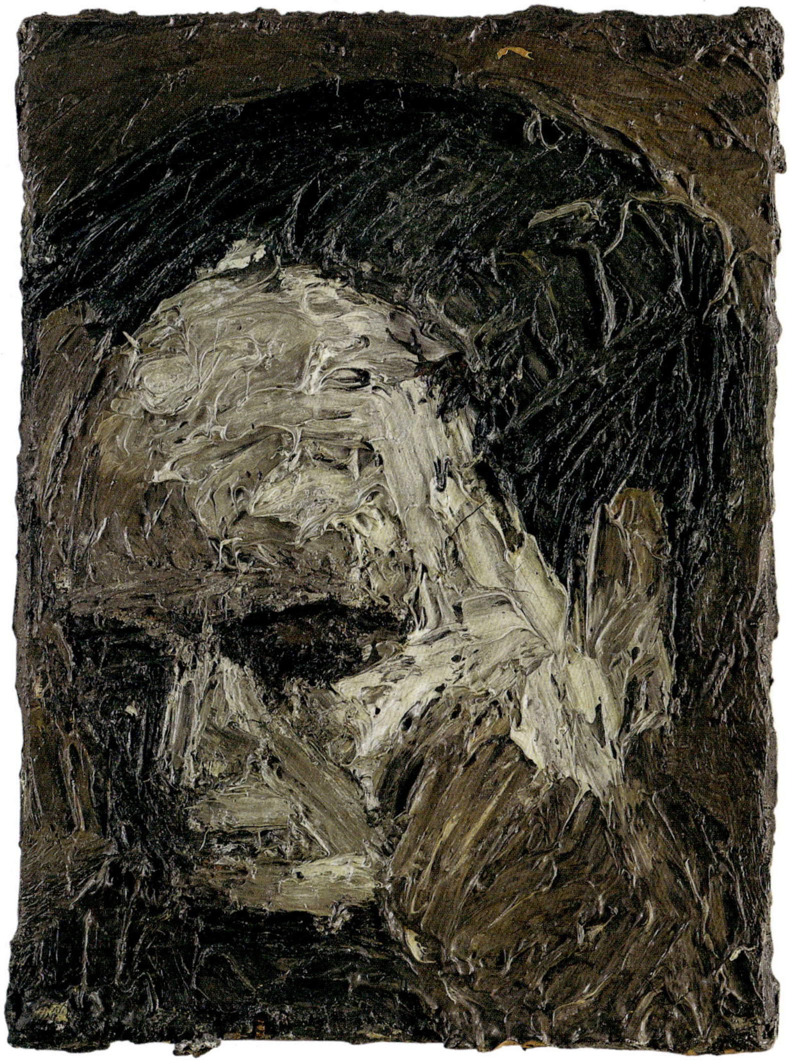

this time. As their friendship developed, from around 1951 they began together to explore the city for suitable motifs. Early subjects included music halls and Smithfield meat market, which Auerbach described as "a warren of passages … I was so excited by it".[5] "London," Kossoff later recalled, was "full of subjects."[6] Both also responded to the visual excitement of the city's many building sites, which they visited together and also independently. One of Kossoff's first subjects was the excavation on the south side of the Thames, which was then being developed for the Festival of Britain. In 1952 he drew building sites in the Oxford Street area. Later he worked in the vicinity of his studio at Mornington Crescent, in the City and around St Paul's, and in proximity to Victoria Street (fig. 30). Like Auerbach, Kossoff's engagement with these subjects continued until around 1962, culminating in that year with his painting *Shell Building Site*.

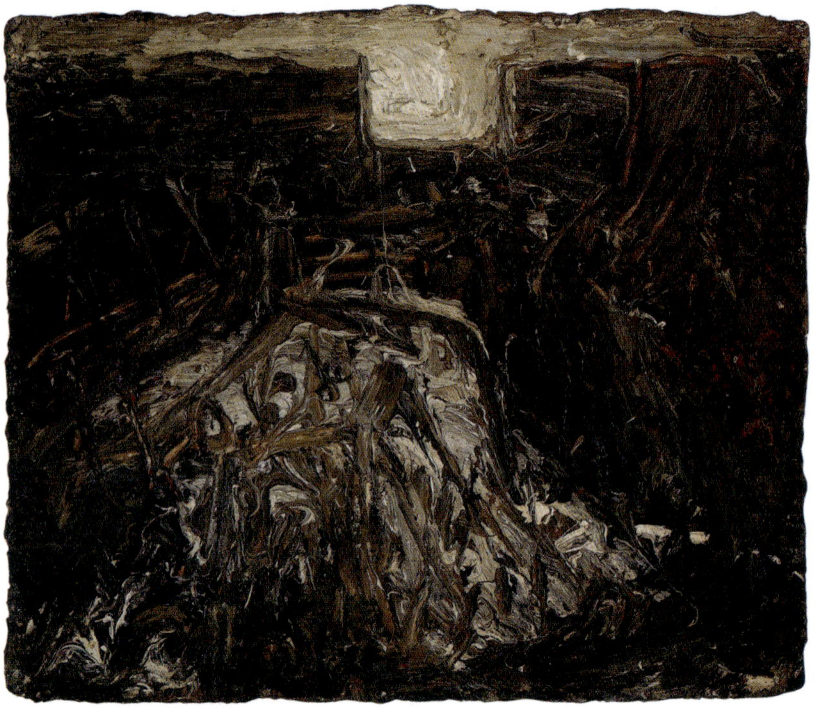

Fig. 30
Leon Kossoff, *City Building Site*
1957
Oil on board, 124.5 × 152.4 cm
Private collection

As they did not always draw together, there is no complete congruence between the sites the two artists painted. Auerbach's earliest motifs were sites on the Earl's Court Road and Bruton Street in 1952–53, and St Pancras and Portobello Road in 1954, none of which appear in Kossoff's work at the time. Auerbach's paintings of a site in Oxford Steet are later than Kossoff's, dating from 1959–61. The subject of his final painting, based on a Leicester Square site in 1962, is unique to him. However, there were also moments when the two young men were engaged with similar subjects. Both appear to have drawn and made paintings of the building site at St Paul's in 1954–55 and of the Shell Building site and the Victoria Street site in the late 1950s.

In part this shared involvement with similar subjects around the same time was due to practical issues. There was a need to know where building work was starting, for a new building site was a short-lived prospect. Once construction commenced a bomb site could quickly become a hive of frenetic activity. This would transform the appearance of the motif in a short space of time. Drawn to the dramatic tableaux that would be generated by figures moving within an large and initially chaotic space, Auerbach and Kossoff shared an imperative to capture the scene while it lasted. For that reason drawing would be carried out energetically, this process contrasting markedly with the more protracted subsequent phase of painting from the drawings in the studio, as the image was slowly digested. The complex changes they witnessed at the sites clearly exercised a fascination for both artists. This spectacle of transformation – involving creation and invention, and the passage from chaos to order – informed each artist's response to their motif.

The Spirit in the Mass

The work of Auerbach and Kossoff would later develop in distinctive ways that are completely independent of each other. Indeed, this divergence can be traced to early differences between their respective responses to building-site subjects. As his *St Paul's Building Site* of 1954 (fig. 31) demonstrates, Kossoff's paintings were characterised by an urgent rush of energy preserved in the tangled facture of the paint. In relative contrast, Auerbach's images, though no less raw, from the outset manifest an instinct for structure. This can be seen, for example, in *Summer Building Site* (1952; cat. 1), in which vector-like lines lock the image in place.

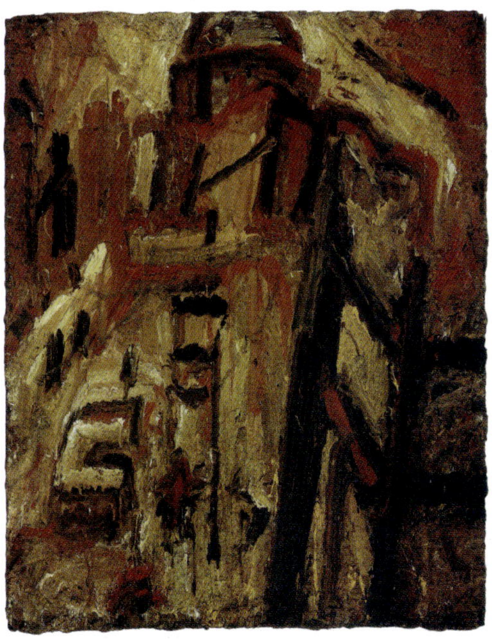

Fig. 31
Leon Kossoff, *St Paul's Building Site*
1954
Oil on board, 152.4 × 121.9 cm
Private collection

At this critical formative stage, however, their respective progress was informed not only by their mutual interaction but also by the example of a shared teacher, David Bomberg. Bomberg's influence on both was profound. Between January and September 1948, before commencing at St Martin's, Auerbach had completed two terms as a temporary student at Borough Polytechnic, where Bomberg taught. After starting at St Martin's in September 1948 he continued to attend Bomberg's evening classes at Borough Polytechnic, twice a week, until 1954. Shortly after meeting Kossoff, Auerbach suggested that he too attend Bomberg's classes, and from 1950 to 1952 the two attended together on this regular, occasional basis, absorbing the older artist's ideas and encouragement.

Both Auerbach and Kossoff appear to have found the training they were receiving at St Martin's constricting. The approach to drawing to which they were exposed was disciplined but yielded results that were somewhat unadventurous – "logical and coherent",[7] in Auerbach's words. In contrast, Bomberg was an intense and inspiring teacher. Kossoff has described seeing Bomberg in action: "Once I watched him draw over a student's drawing. I saw the flow of form, saw the likeness to the sitter appear. It seemed an encounter with what was already there and I'll never forget it."[8] At the heart of this approach was an insistence that drawing had to be more than an attempt to make a literal reproduction of the appearance of things. In particular, Bomberg disliked drawing which depended on facility without feeling. For him, drawing should combine close observation with intuition and expression, an approach he described memorably as a quest for the "spirit in the mass". Such an approach eschewed mimetic illustration in favour of an image arrived at instinctively, one that evoked the observed subject through the creation of marks and gestures that conveyed the apprehension of its mass, weight, structure and presence. "The eye," Bomberg claimed, "is a stupid organ."[9]

For that reason the creation of an image involved drawing upon, and finding equivalents for, a level of internalised experience that was not simply, or only, visual.

As Bomberg's students were led to understand, the experience of *matter* was a paramount concern and in the endeavour to render matter formally the idea of 'actuality' was central. The invented image should convey, by equivalent means, the physical reality of the subject as grasped by the mind of the observer, its independence as a tangible thing occupying space, anchored by gravity. As Auerbach later described, while the process of creating such an image was essentially intuitive, this objective would not be won by quick solutions: "He had this sort of idiom that allowed one to go for the essence at the very beginning, to adumbrate a figure in ten minutes and then to re-do it and then to find different terms in which to re-state it until one got something, however unlike a poster of a figure or a photograph, that seemed to contain the mind's grasp of its understanding of its subject .... This wasn't a question of display, this was a question of a private quest which had certain results."[10] Viewed in the light of these words, *Summer Building Site* (1952) marks an important advance in Auerbach's development because, with this first building-site painting, there is clear evidence of his success in putting into practice the example set by Bomberg.

The painting came at a watershed in Auerbach's studies. He had spent three years studying at St Martin's, concurrent with his attendance at Bomberg's evening classes. With the prospect of transferring to the Royal College of Art in sight he spent the summer of 1952 working on the painting, using drawings made at the outset on site. Throughout this period he remained wedded to a naturalistic approach while employing a conventional palette. As the summer wore on he felt increasing frustration at his inability to break with habits that he no longer trusted. Such practices now sat ill with the ideas to which he had been exposed in Bomberg's classes. According to Auerbach's own account, the imminent move and the prospect of further study appear to have brought matters to a head. At the end of the first day of the new term he returned home and took a radical decision. Effacing his earlier efforts, he reworked the painting entirely.

The result is a triumph of adumbration and concentrated formal rigour. In place of an assembly of individual descriptive details, the image has been distilled to its essential, underlying structure. The composition reads as a unified whole in which flattened planes, extended vectors and strongly contrasted areas of colour contain intimations of space, depth and recession. The painting relates to its subject not through illusion and imitation but as a plastic equivalent. The image is experienced as a tangible object in its own right, defined by its own internal geometry, balance and formal rhythms. The observed world that was its source has been reinvented and re-stated as a new, visual fact. The way that, overall, the painting has been drawn into a single coherent unity yields an impression of a perceived subject having been grasped, digested, and then reduced to its essence. In this respect the painting intimates its relationship with Bomberg's teaching. At the same time, however, *Summer Building Site* is recognisably the authentic expression of a new, personal voice, one unique to Auerbach himself.

Eight years earlier, in 1944, Bomberg had also made drawings of London in the wake of wartime bombing. His related painting *Evening in the City of London* (1944) presents a poignant view of the devastated city with the dome of St Paul's its focal point, rising from a ragged horizon (fig. 32). However, whereas Bomberg's image presents a panoramic view looking across the city, Auerbach's response to London's destruction is altogether different (and, indeed, there is no evidence that he was aware of Bomberg's earlier painting). In making drawings on site, Auerbach would be in close proximity to the subject, sometimes having climbed mountains of excavated earth in order to gain a vertiginous vantage-point. This view characterises *Summer Building Site*, in which a huge pit opens up the foreground. This is traversed by a vector-like line suggesting a plank along which workmen would make a precarious journey. Auerbach's choice of subject – an artificially created precipice – was

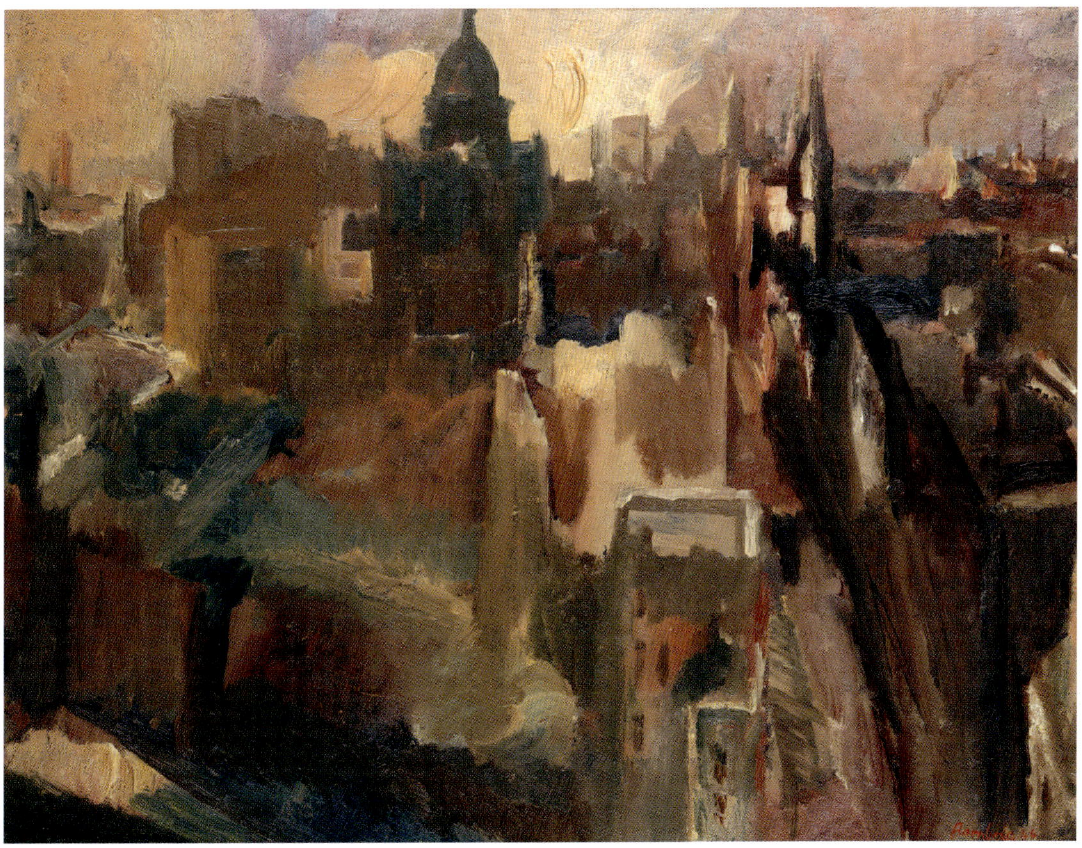

Fig. 32
David Bomberg, *Evening in the City of London*
1944
Oil on canvas, 69.8 x 90.8 cm
Museum of London

entirely new, as was his treatment of the motif, involving a view looking downwards into a cavernous hole. It is difficult, too, not to sense an affinity between the subject and Auerbach's own situation, which was not simply one of physical proximity. The view presented leads the eye into and across an uncertain space – unknown terrain which he would now explore employing the new visual language at his disposal.

The paintings of building-site subjects that Auerbach made between 1953 and 1955 mark both a continuation and a radical extension of the ground opened up by *Summer Building Site. Building Site, Bruton Street* (1953; cat. 2), which followed, advanced a new palette of dark earth colours in which image, texture and shape appear to have coagulated. The reality of the motif has been transmuted, so that the resulting painted image has a corporeal character that can be sensed as well as seen. The viewer 'feels' the weight, texture and substance of the subject through eyes that call upon the memory of touch.

As these paintings proceeded, increasingly the distinctive character of Auerbach's endeavour emerged. With each succeeding work the physical substance of the image deepened. *Building Site, Earls Court Road* (1953; cat. 3) is pitted and rutted like a patch of sun-bleached earth. *Building Site, St Pancras* (1954; cat. 4) bears the channels and grooves of movement enacted while the paint remained pliable. But such qualities are not confined to the surface, nor is the thickness of the paint an end in itself. Image and material density are concomitant and inseparable. The painting is experienced as a material fact in which meaning and reference are embedded. In this way the paintings are at one with their subjects, almost as if formed from the excavated earth, mud and debris that inspired them.

However, it would be a mistake to view the density of the paint only as an analogue for the material fabric of the observed scene. Auerbach has commented: "The sense of corporeal reality, that's what matters". In these words there is an acknowledgement of congruence between the paintings' physical form and their immediate subject. At the same time there is an implication not only of mute physical substance but, beyond that, of the reality that is specific to the experience of inhabiting a living body. The intimation of an equation between the form of the painting and human experience is a vital consideration, not only in terms of the process by which the building-site paintings came into being, but also in relation to the character and meaning of these works.

Existence precedes essence

In common with all Auerbach's building-site paintings, *Building Site, Bruton Street* was based on earlier drawings made on site. One such pencil drawing, *Study for Building Site, Bruton Street* (1953; cat. 2 (i)), provides an illuminating comparison with the resulting painting. The drawing is precise and literal, even laboured, in its rendering of detail. In contrast, the related painting is almost a complete reinvention in which individual features have been absorbed into a broader physical mass.

Building Site, Bruton Street extends those earlier developments present in *Summer Building Site*. As with the earlier painting, in its initial stages the image proceeded conventionally, employing a range of thinly applied colour in imitation of the drawing. Again at a late stage, the entire image was repainted using earth colours, a radical transformation which subsumed individual details within an overall formal reorganisation.

Repeating this way of working appears to have confirmed what now became a consistent process. Drawing *in situ* would be followed by a protracted period of studio-based painting during which the source image would be progressively stated and re-stated, each re-creation transforming the preceding state of the image. As a result of this ongoing improvisation, the work would grow towards an unknown destination. As it proceeded the image would acquire a deepening physical mass through the accretion of paint. At the same time, through an accumulation of choices and decisions, a painting would emerge that would be less a record of the original motif than something entirely new, forged from the sum of these actions.

This process of progressively restating an image bears a clear debt to Bomberg's quest for the *essence* of the subject, the "spirit in the mass". But the idea of seeking essence through the artist's complete involvement was characterised by Auerbach as an "experimental journey" and as "a private quest".[11] These phrases are pregnant with the implication of personal discovery. The search for essence should therefore also be seen in relation to the widespread circulation of existentialist theories in London's art schools, bars and cafés from the early 1950s. The writings of Jean-Paul Sartre, Albert Camus and Simone de Beauvoir identified a world in which man existed in the absence of any divine plan, with no ordained nature or purpose. As Sartre expressed it, "He is nothing".[12]

Within a wider intellectual debate, Sartre's *Being and Nothingness: An Essay on Phenomenological Ontology* (1943) and his statement of doctrine *Existentialism and Humanism* (1946) are key texts, adumbrating a universal predicament and proposing existentialism as a philosophy that made human life possible. Taking as his premise that man must forge his own nature and purpose, Sartre declared that man is "what he makes of himself".[13] Through his decisions and actions man defines his essential nature, an evolving process that is renewed constantly. This idea is the basis of existentialism's first principle, as stated by Sartre, "that existence precedes essence, or, if you prefer, that subjectivity must be the starting point".[14] The primary role of subjective experience in defining human nature was also affirmed by Maurice Merleau-Ponty in his *Phenomenology of Perception* (1945). Merleau-Ponty described the world as a field of experience, the human body being a site for impressions inspired by external sources which were interpreted subjectively. Such ideas identified the fabric of the world, of existence itself, with human experience, a concept that Sartre encapsulated in the observation: "There is no universe other than a human universe, the universe of human subjectivity".[15]

In Paris, such texts, and the further dissemination of existentialist ideas through public lectures, plays and journalism, led to a rapid, widespread popular fascination with the doctrine. Inevitably artists such as Giacometti, Dubuffet, Fautrier and Wols were caught in its spreading appeal, their work being seen as evidence of the philosophy finding authentic artistic expression. By the early 1950s, the language and ideas of existentialism had spread to London, acquiring currency as a new cultural force generating debate and critical discourse. In particular, around this time the Royal College of Art became a focus for discussion along existentialist lines about the future direction and development of realist painting, not least as a result of David Sylvester's lecture, 'Towards a New Realism', which he delivered there in 1951. Sylvester had lived in Paris from 1947 to 1950 and returned to London armed with close, first-hand acquaintance with the tenets of existentialism he had encountered abroad. The premium his lecture placed on the role of experience and the importance of active involvement bore a clear relation to Sartre's ideas concerning the primacy of subjectivity. In transferring to the Royal College of Art in 1952, Auerbach had immediate contact with the intellectual atmosphere generated by Sylvester's lecture and by the presence of Francis Bacon, who had a studio at the college from 1951 to autumn 1953.

Beyond the College, existentialist ideas gradually spread, colouring the behaviour and attitudes of a wider group with no first-hand contact with the doctrine's theoretical base. On 3 August 1955, the first English-language production of Samuel Beckett's play *Waiting for Godot* opened at the Arts Theatre Club in London and was described by one critic as a perfect example of an existentialist play.[16] Its cast of tramps and misfits and their absurd efforts to pass time in a bleak landscape at first elicited incomprehension. Its subsequent popular success, and the discussion that raged in the Press in early 1956, were a demonstration that existentialism had become an essential part of the fabric of the time. Indeed by 1957, Sartre could write with some justification that the word 'existentialism' "is all the rage".[17]

If these intellectual and cultural events formed the backdrop against which Auerbach's building-site paintings developed, this is not to say that he was responding directly to the ideas formulated by existentialism. But they gained such currency during this period that not to see Auerbach's arresting images of dereliction and creation in these terms would be perverse.[18] Certainly the emphasis placed by existentialism on the central role of choice and the idea of total involvement are germane to an appreciation of Auerbach's work. Sartre's claim that "subjectivity of the individual is ... our point of departure"[19] goes to the heart of an artistic activity rooted in the primacy of private experience. Of his work at this time Auerbach later commented: "You've got to venture into unknown territory where you're trying to state the thing, without having these hand-holds and grips and assistance of previous practice and so when the paintings became these strange lumps of thick paint, I was very interested in that and I pursued that line".[20]

The paintings of building-site subjects that Auerbach completed between 1955 and 1962 are essentially products of this will to continue to explore uncharted terrain, a context at once physical and artistic. During this period his motifs included sites at St Paul's and at Hungerford Bridge, the South Bank viewed from the Festival Hall and the Thames, Oxford Street, Victoria Street, Maple Street and Leicester Square. As if responding to an emerging order both in the observed landscape and within his own practice, in this progression of paintings there is physical evidence of growing authority and an imposing pictorial presence commensurate with the drama of his subject-matter. The related pair of *Oxford Street Building Site* paintings (cat. 11a and 11b) and the *Maples Demolition* painting (cat. 12) are all bathed in a warm, earthy glow, long vector-like lines and other structural elements securing the image within the densely worked facture. The evident confidence of these later paintings of building sites manifests a new coherence, one arising from a long process of making and formation (see fig. 33).

Sartre made a telling connection between the way man must define his own nature by his actions and the predicament of the artist confronted with an unknown outcome: "It is clearly understood that there is no definite painting to be made, that the artist is engaged in the making of his painting, and that the painting to be made is precisely the painting he will have made. It is clearly understood that there are no *a priori* aesthetic values, but that there are values which appear subsequently in the coherence of the painting."[21]

Like their subject matter, Auerbach's building-site paintings emerged from a kind of inchoate matter through a protracted process of making. But there the analogy with the subject ends. Unlike the process of building, the artist's procedure involved choosing and inventing, repeated deliberation, statement, dissatisfaction, despair, changes of direction and restatement. As Auerbach has explained, "All my paintings are the result of hundreds of transmutations".[22] Viewed in this way, these early paintings established an outlook which was rooted not only in depiction. Their form and expressive content were the result of the artist's accumulated actions, impressed with his individual perceptions, driven by a personal will to continue. As Sartre observed, "The feeling is formed by the acts one performs".[23]

Fig. 33
Frank Auerbach in his studio, 1964
Photograph by Snowdon
Camera Press, London

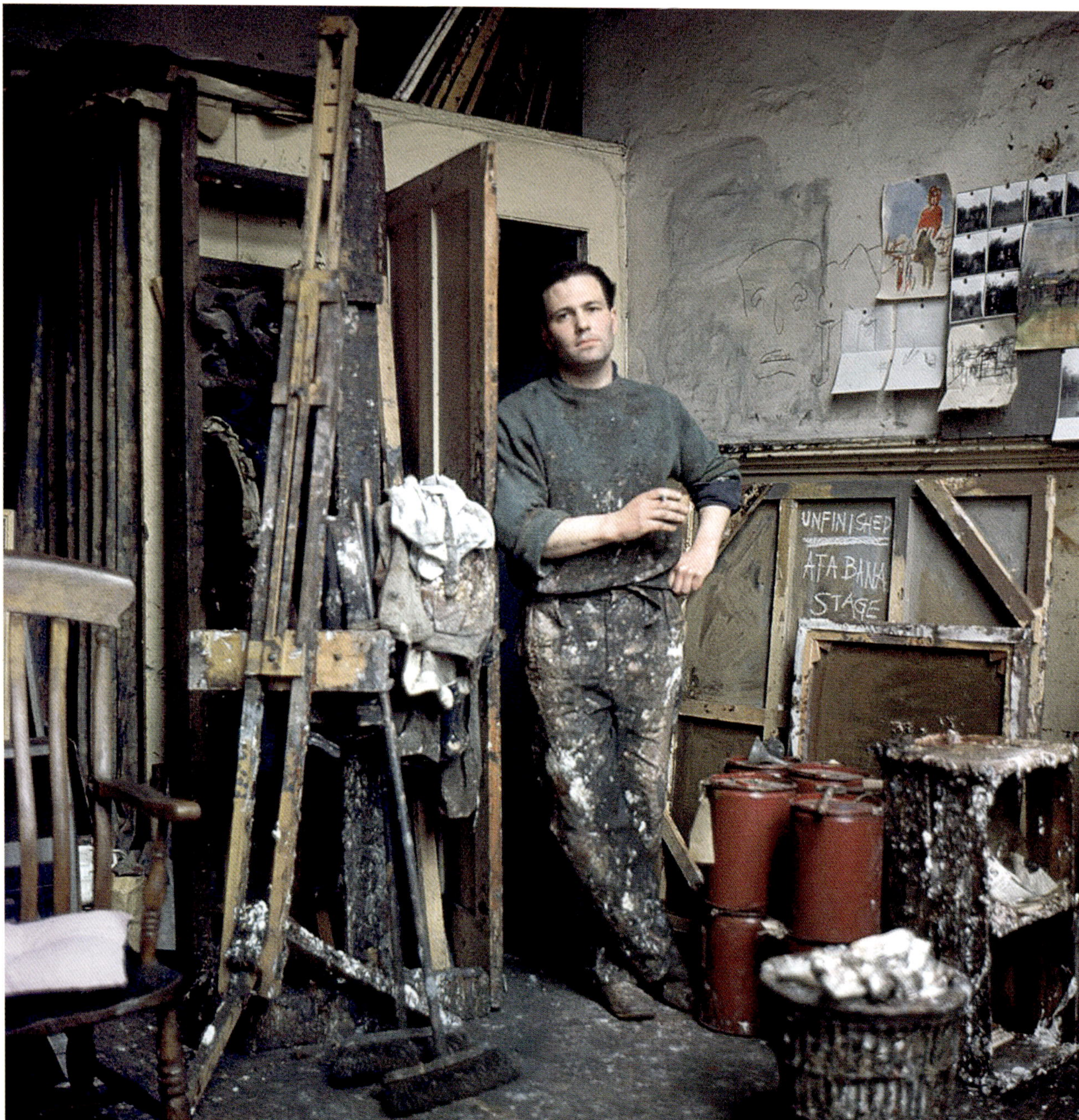

Arising from the immediate situation he witnessed, a gradual process of renewal wrested from the ruins of war, Auerbach's building-site paintings are a remarkable document of a unique moment. But the particular nature of Auerbach's response, in which painting can be seen defining itself afresh, invests this body of work with singular significance. In his later comment, "Nothing can be left out ... you have to bury the irrelevant in the picture somehow",[24] there is an echo of Samuel Beckett's observation, "There's always an excess of circumstance".[25] Both these assertions affirm the value and fragility of individual experience. Auerbach's building-site paintings are themselves the physical embodiment of human endeavour, a vast ensemble of actions directed towards "a virgin future",[26] as Sartre described it. They give lasting form to that which is ephemeral, fleeting and, in human terms, vital. As such they have an unexpected poignancy – and a grandeur.

Notes

1 Auerbach/Wright interview 2009.
2 Auerbach/Tusa interview 2009.
3 Auerbach/Wright interview 2009.
4 Quoted in *The Forgotten Fifties*, (exhibition catalogue), Graves Art Gallery, Sheffield (and tour), 1984, p. 49.
5 Auerbach/Wright interview 2009.
6 Unpublished notes of interview with Leon Kossoff by Paul Moorhouse, July 1995.
7 Quoted in Robert Hughes, *Frank Auerbach*, London, 1990, p. 30.
8 Quoted in Richard Cork, 'City of the Heart', *Times Magazine*, 27 May 1995, p. 27.
9 Quoted in Hughes 1990, p. 31.
10 Auerbach/Tusa interview 2001.
11 *Ibid.*
12 Jean-Paul Sartre, 'Existentialism and Human Emotions' (1957), reprinted in Gordon Marino (ed.), *Basic Writings of Existentialism*, New York (The Modern Library) 2004, p. 345.
13 *Ibid.*
14 *Ibid.*, p. 344.
15 *Ibid.*, p. 366.
16 Katharine Wilson, quoted in James Knowlson, *Damned to Fame – The Life of Samuel Beckett*, London (Bloomsbury) 1996, p. 416.
17 Sartre, 'Existentialism and Human Emotions' (as note 12), p. 343.
18 Such connections were made early on, for example in David Wright and Patrick Swift (eds.), *X: A Quarterly Review*, vol. 1, no. 1, London (Barrie and Rockliff) 1959, which reproduced Auerbach's 'Fragments from a Conversation' and reproductions of his paintings next to Beckett's text 'L'image' and drawings and a text, 'The Dream , the Sphinx and the Death of T', by Alberto Giacometti.
19 *Ibid.*, p. 357.
20 Auerbach/Tusa interview.
21 Sartre, 'Existentialism and Human Emotions' (as note 12), p. 361.
22 Quoted in Norman Rosenthal, 'Auerbach and His History', in *Frank Auerbach: Paintings and Drawings 1954–2001*, (exhibition catalogue), Royal Academy of Arts, London, 2001, p. 26.
23 Sartre, 'Existentialism and Human Emotions' (as note 12), p. 352.
24 Hughes 1990, p. 135.
25 Quoted in Catherine Lampert (ed.), *Michael Andrews* (exhibition catalogue), Arts Council of Great Britain, London, 1981, p. 53.
26 Sartre, 'Existentialism and Human Emotions' (as note 12), p. 350.

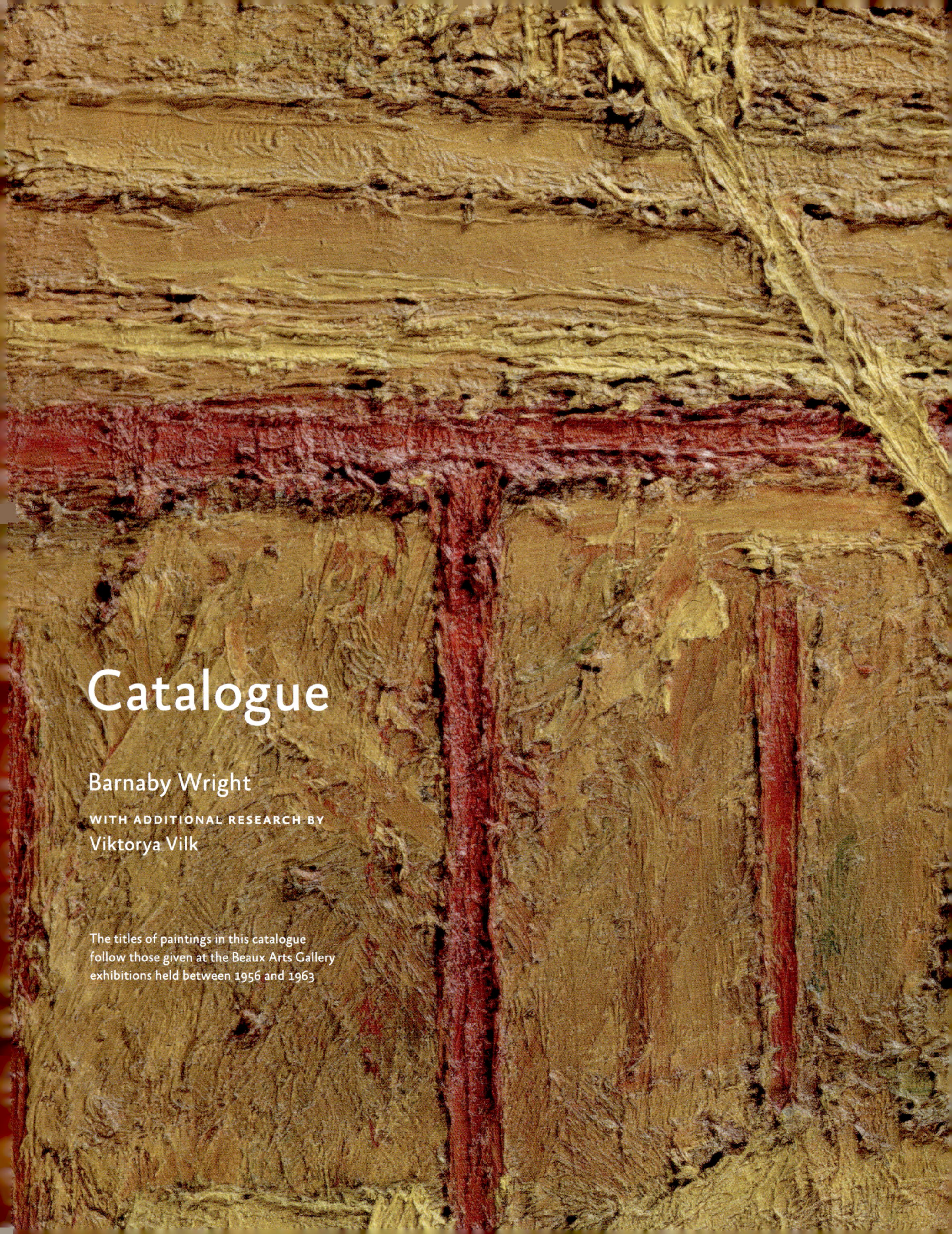

Catalogue

Barnaby Wright

WITH ADDITIONAL RESEARCH BY
Viktorya Vilk

The titles of paintings in this catalogue follow those given at the Beaux Arts Gallery exhibitions held between 1956 and 1963

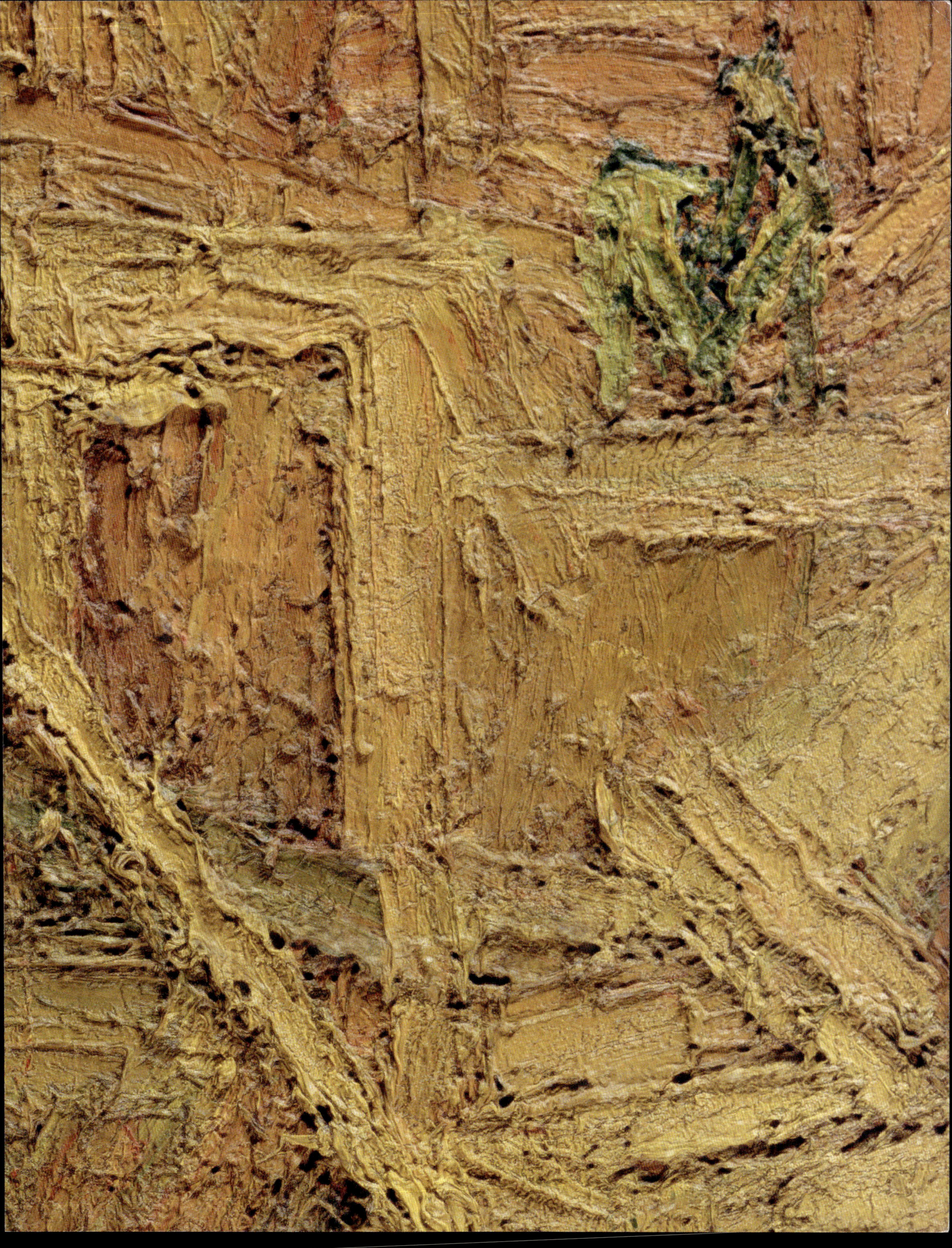

1

Summer Building Site

1952

Oil on board
76.2 × 106.7 cm
Private collection

SELECTED EXHIBITIONS

'Degree Exhibition', Royal College of Art, London, 1955; *Frank Auerbach*, Beaux Arts Gallery, London, 3–28 January 1956 (not listed in published catalogue but added as a hand-written amendment to the Beaux Arts Gallery's copy); *Contemporary British Landscape*, Arts Council, Glynn Vivian Art Gallery, Swansea, 12 November–3 December 1960 (and nationwide tour); *Helen Lessore and the Beaux Arts Gallery*, Marlborough Fine Art, London, February 1968; *Frank Auerbach*, Hayward Gallery, London, 4 May–2 July 1978, and Fruit Market Gallery, Edinburgh, 15 July–12 August 1978

Summer Building Site was a breakthrough painting for Auerbach. "I felt it was the beginning of my life as a painter," he recalled recently.[1] It was the first work in which the twenty-one-year-old artist thought he had found his own artistic voice. It would also prove to be the first of what would become over the next decade a group of fourteen London building-site paintings.[2]

The subject was a building site on the Earl's Court Road, close to Kensington Police Station.[3] Auerbach had often passed it on his way to visit Estella West, his frequent model, who had a house nearby at no. 81, where Auerbach had previously lodged.[4] He probably began drawings of the site at the beginning of summer 1952 (none of these appear to survive). He recalled starting the painting in "quasi-naturalistic" colours and labouring over it for many weeks with increasing dissatisfaction.[5] At the beginning of September he reluctantly took up a place at the Royal College of Art, disillusioned by the prospect of entering yet another art school after several years of previous study. He remembers returning home from his first day there in a temper and taking up *Summer Building Site* again, repainting "the whole thing from top to bottom . . . totally disrupting it" with bright oranges and yellows.[6] Through his working and reworking of the image the scale of individual elements became distorted, breaking his previous fidelity to a naturalistic rendering. "Suddenly I was conscious of something underneath it," he wrote in 1959, "I'd destroyed all the reminders (that is of painting) to get a unique thing . . . it began to operate by its own laws."[7] David Bomberg's teaching at the Borough Polytechnic had inspired Auerbach to work in this persistent manner, exploring his subject from all angles, to seek out what Bomberg called "the spirit in the mass".

In the centre are two trestle ladders supporting a plank and the small red figure of a workman, perhaps bending to lay bricks, with a larger figure to the right. These simplified forms of workmen and building equipment offer an emblematic image of the construction site. By contrast the prominent yellowish lines which form a v-shape running across the painting are less clearly representational. This vectoring line may be suggestive of scaffolding or a beam spanning the void of an excavation, but it is also assertively a piece of pure painting which functions in the construction of the composition itself. In this way Auerbach established a sense of empathy between his own labours with paint and the imagery of the building site, which would become an enduring feature of his subsequent works.

Summer Building Site is experimental, and distinct from others in the group. Its impasto is much thinner than the extraordinarily thick paint surfaces that would become a defining feature of subsequent works. However, its exploration of the tensions between line and form anticipates later paintings (see for example cat. 12 and 13). Already in *Summer Building Site* Auerbach achieves a sense of the struggle between raw earth and architectural order which he found so compelling in the construction sites of post-war London.

NOTES

1 Auerbach/Wright interview 2009.
2 Auerbach recalled having painted "one or two building sites" prior to *Summer Building Site* whilst at St Martin's College of Art, which he destroyed "because they weren't good enough", Auerbach/Wright interview 2009. He is likely, prior to 1952, to have made sketches of various London building sites which were never realised as paintings.
3 The precise location is unknown.
4 At this time Auerbach was living on Anselm Road near Fulham Broadway. It was here that he painted this work and cat. 2 and 3.
5 Auerbach/Wright interview 2009.
6 Auerbach/Wright interview 2009.
7 Auerbach 1959, p. 32.

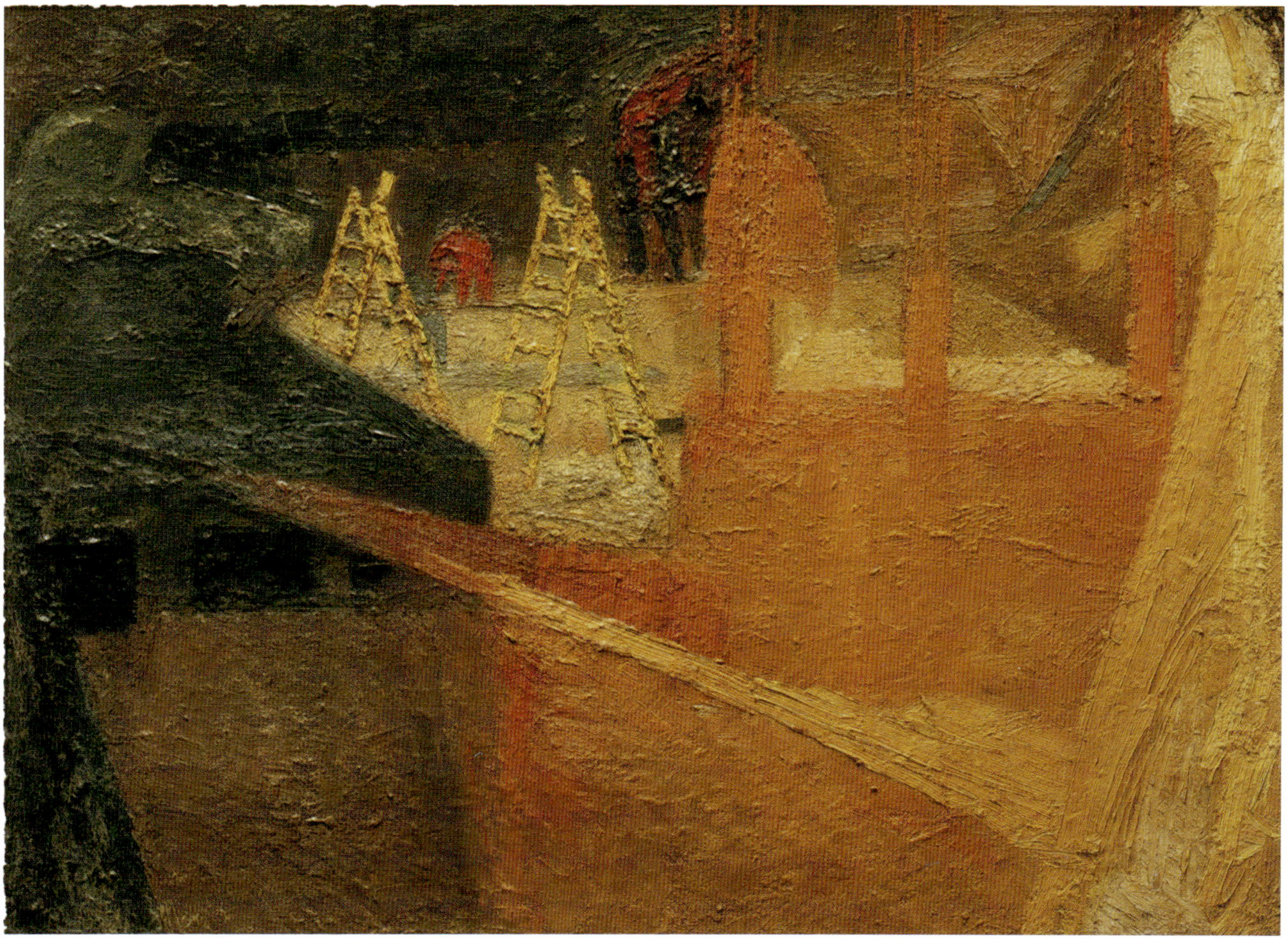

2

Building Site, Bruton Street: winter

1953

Oil on board
76 × 100.2 cm
New Walk Museum and Art Gallery, Leicester

2 (i)

Study for Building Site, Bruton Street: winter

c. 1952–53

Pen and ink, watercolour and pastel (squared up in green pastel and numbered in pencil)
59 × 76 cm
Private collection

SELECTED EXHIBITIONS

(*Building Site, Bruton Street: winter*): 'Degree Exhibition', Royal College of Art, London, 1955; *Frank Auerbach*, Beaux Arts Gallery, London, 3–28 January 1956

All of Auerbach's building-site paintings developed from numerous sketches (most of which he subsequently destroyed), made on the various sites. However, this is the only work from the group for which he produced such a large-scale preparatory drawing, which itself would have been worked up in the studio from smaller on-site sketches and was then squared up in green pastel for transfer on to the primed board.[1] A comparison of the drawing and painting offers a fascinating insight into Auerbach's working process during this crucial phase in his artistic development. Between 1952 and 1953 the young artist was realising that he could only achieve a satisfactory form of expression by painting and repainting his subjects over many months, resulting in the gradual build-up of layer upon layer of thick paint. He recalled beginning this work in fairly thin strokes of blue, purple, green and yellow paint, retaining much of the detail seen in the drawing.[2] Finding this "dead and bitty", he repainted it more freely, beginning a process of slow alteration and distortion of the original composition. Spurred on by John Minton, one of his tutors at the Royal College of Art, Auerbach began reworking it more radically and repainted the whole composition in monochromatic and earth tones. This continual reworking of the initial image was perhaps, he explained, "an attempt to find a secret internal geometry that lay behind it, the core as it were".[3] It has remained an enduring feature of his painting technique to this day.[4]

Initially, the finished painting seems almost unrecognisable in relation to the original drawing, with much of the clarity of detail buried beneath successive paint layers. Closer examination reveals that the prominent light-coloured triangular form in the foreground is in fact an enlarged and reshaped version of the mound of sand seen in the drawing. The similarly coloured diagonal strip leaning against the left-hand edge of the painting is a blockish rendering of the thinly drawn ladders. Further scrutiny reveals that the steel-framed grid of the building's skeleton, so prominent in the drawing, is partly obscured in the painting by the dark paint-surface of the background. The painting reverses the compositional priorities of the drawing. The amorphous character of Auerbach's paint threatens to subsume the structuring geometric grid of the building, while the ladder and mound of sand, which appear incidental in the drawing, are enlarged and reformed to become the dominant features of the composition. The three-dimensional rendering of the building and the black-and-white post, which give a sense of depth and balance to the drawing, are largely eliminated from the painting, resulting in a shallower picture-plane. This enables Auerbach to give dramatic expression to the immediacy of the paint surface, which is not kept fully in check by perspectival devices and appears to press itself out of the frame.

Although Auerbach identified the location only as Bruton Street in his title, in fact the construction site was that of the Time–Life Building on the corner of Bruton Street and New Bond Street (fig. 18). It was designed by the architect Michael Rosenauer and built between 1951 and 1953.

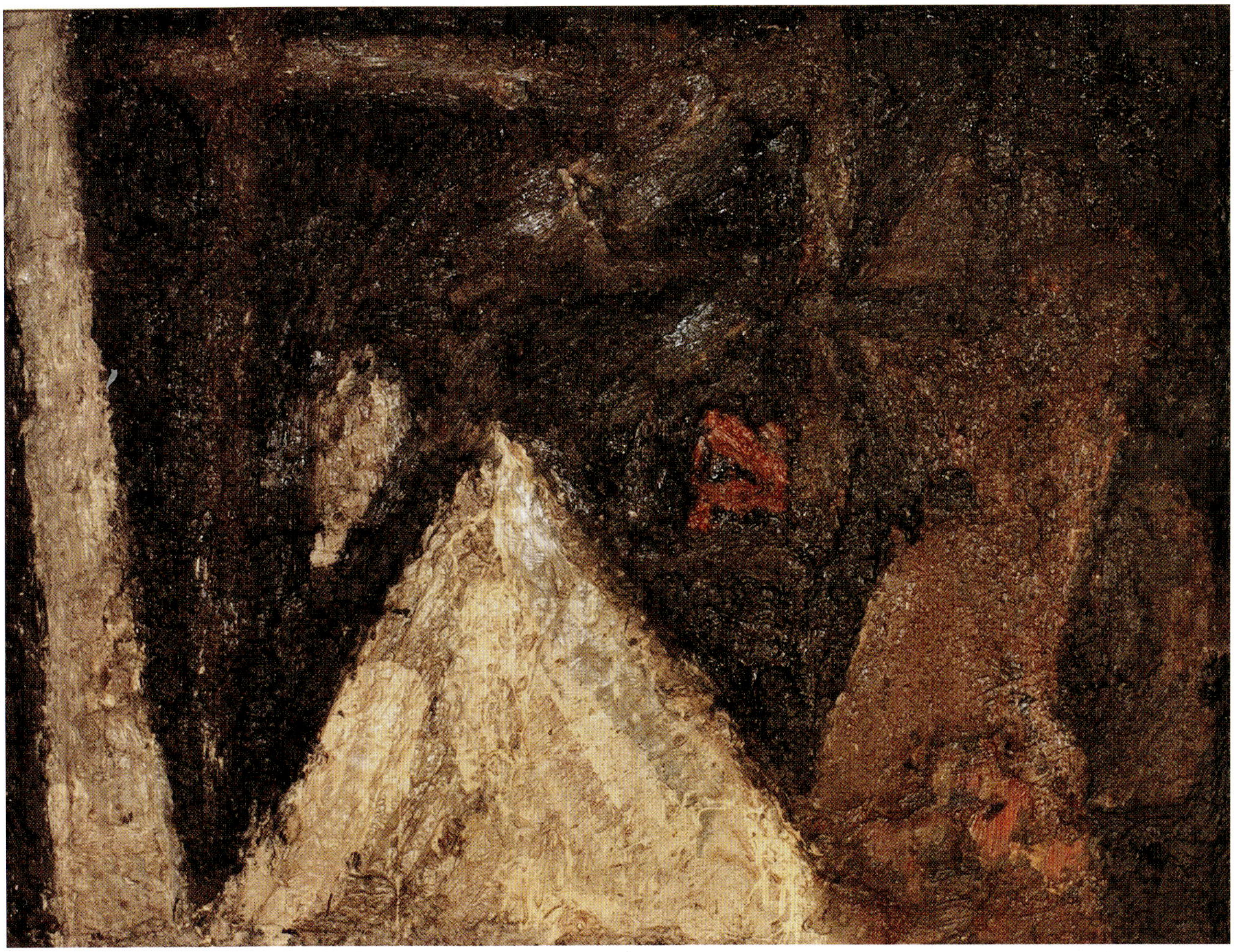

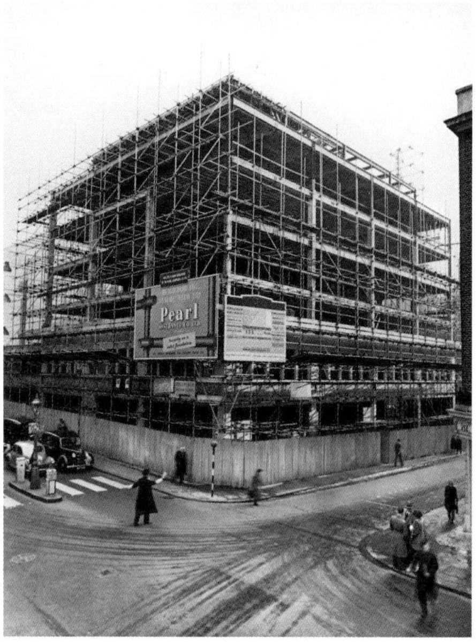

Fig. 34
The Time–Life Building under construction on the corner of Bruton Street and New Bond Street, 1952
Photograph: Hulton Archive/Getty Images

It opened to considerable publicity and critical interest in the Press. It was the first post-war company headquarters in London conceived in a Modernist architectural style and therefore contributed a powerful statement to the debates surrounding the post-war reconstruction of the city. Generously financed by American capital, the building boasted lavish use of Travertine marble and Portland stone. The building quality appeared especially extravagant in Britain during this period of post-war austerity and emphasised the economic power of the United States. The project was conceived as a showcase for British artistic and design talent. Sir Hugh Casson, who had been Director of Architecture for the 1951 Festival of Britain, was appointed to commission artists and interior designers to contribute to the building. Nowhere was this demonstrated more visibly and powerfully than in the building's exterior sculptural screen by Henry Moore, who had emerged after the War as Britain's foremost modern sculptor.

Auerbach made his drawings at a relatively early stage in the building's construction, perhaps in late 1951 or early 1952, when only three or four stories of the steel frame seem to have been in place[5] and well before the scaffolding was erected around the frame to begin the exterior marble and stone cladding work, as seen in a 1952 photograph (fig. 34). This typified the artist's preference for building sites in the early phases of their construction and at the point when architectural order had not yet been fully imposed upon the apparent chaos and formlessness of the site. In fact, the Time-Life Building was relatively well advanced by Auerbach's standards at the point at which he drew it and it is telling that in the transition from drawing to finished painting he should so thoroughly have challenged the dominance of the building's Modernist grid-pattern structure.

NOTES

1 Auerbach also produced an oil sketch on paper for this work, which now appears to be lost along with the other pencil sketches: Auerbach/Wright interview 2009. Auerbach recalled that much to his fury the Royal College of Art's carpenter cut a small strip off the left-hand edge of the painting so that it would fit a pre-existing frame.
2 Spatters of yellow oil-paint can be seen on the right-hand side of the drawing.
3 Auerbach/Wright interview 2009.
4 A crucial difference is the fact that, since the mid 1960s, Auerbach has scraped down his painted surfaces before reworking them rather than building up layer upon layer of paint.
5 The verso of *Study for Building Site, Bruton Street* is signed and dated 1950. It seems likely that this date was added much later and that Auerbach misremembered the precise date.

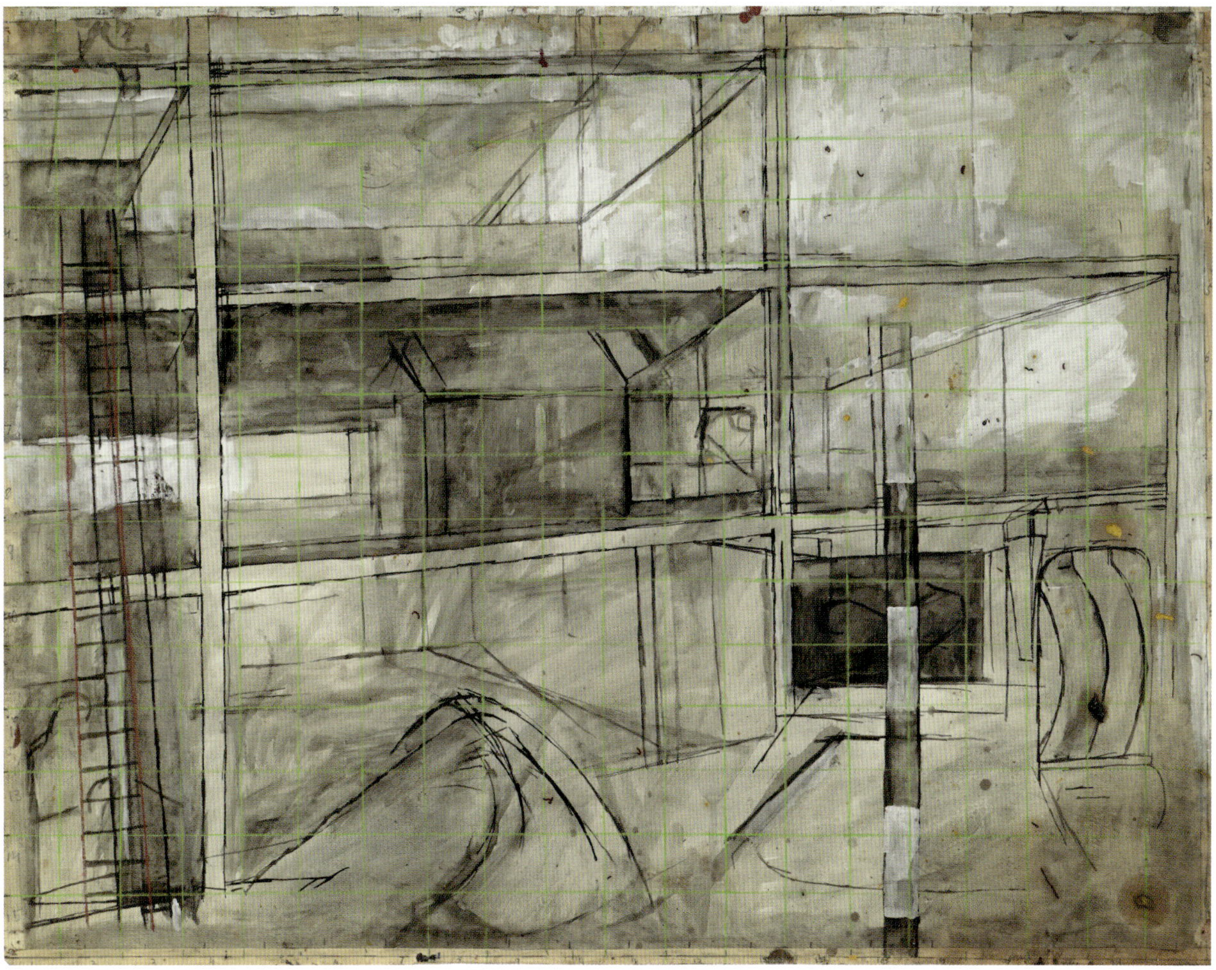

2 (i)

3

Building Site, Earls Court Road: winter

1953

Oil on board
91.5 × 122 cm
Private collection

SELECTED EXHIBITIONS

'Degree Exhibition', Royal College of Art, London, 1955; *Frank Auerbach*, Beaux Arts Gallery, London, 3–28 January 1956; *Frank Auerbach*, Hayward Gallery, London, 4 May–2 July 1978, and Fruit Market Gallery, Edinburgh, 15 July–12 August 1978; *The Forgotten Fifties*; Graves Art Gallery, Sheffield, 31 March–13 May 1984 (and tour); *Transition: The London Art Scene in the Fifties*, Barbican Art Galleries, London, 31 January–14 April 2002

In 2001 Auerbach was asked what advice he would give a young art student. "Find out what matters most to you and pursue it", he suggested, and make sure "to try and record it in an idiom that is your own, and not to give a damn about what anybody else says to you".[1] When asked which of his paintings came close to matching these criteria, he singled out *Building Site, Earls Court Road*, produced whilst still a student himself at the Royal College of Art.

"[It] is just a few very thick lumps of black and white and red ochre and everything, all the experience of making the painting which was actually very elaborate at many stages is buried there, and it seems to me to be a sort of remarkable object, and though nobody else may be aware of it, I am aware of the amount of painting experience that's buried under those heavy lumps of black and white and ochre."[2]

The painting is one of the most uncompromising of Auerbach's building-site series. Its subject-matter is barely legible beyond some basic forms and a red element of what appears to be scaffolding towards the centre. However, the work was the result of a particularly lengthy and complex process of creation. Auerbach began painting from sketches and more elaborated drawings of the site (which do not survive) before introducing some tools from his studio into the composition.[3] "I actually posed a still life in the front, of a saw and a pair of pinchers and a hammer, and thought I would create some sort of connection between painting from life and painting from drawings," he recalled. "It seemed to me that certain pictures that I admired had done this and that the conflict between the slight sheerness and blandness that comes in working from drawings and the awkwardness and oddity that the objects in front of one have might make a more convincing image."[4] Auerbach's restless reworking of the subject resulted in expanses of thickly massed paint, its skin dried to a deeply wrinkled and puckered surface. It is as if the paint has consumed almost all vestiges of the image itself.

The work was one of six building-site paintings included in Auerbach's debut solo exhibition at the Beaux Arts Gallery in January 1956.[5] His thickly built-up paint-surfaces presented a considerable and unexpected challenge to most reviewers. Some dismissed the works as "eccentric" or simply "clumsy". But several others recognised that Auerbach had, as David Sylvester put it, "extended the power of paint to re-make reality".[6] Given subsequent insights into the complex creation of *Building Site, Earls Court Road*, John Russell's observation seems particularly relevant: "The battle of representation is fought out a good inch below the topmost peaks of the paint".[7]

The painting's importance to Auerbach is further suggested by the fact that he appears with it in one of the few early photographs taken in his studio (frontispiece). He also produced a more thinly painted replica of the work to present to the Royal College of Art as his graduation piece (fig. 35).

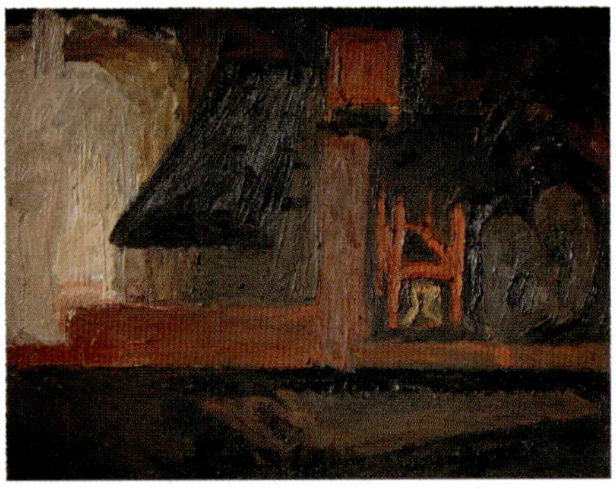

Fig. 35
Frank Auerbach, Replica of *Building Site, Earls Court Road: winter*
1955
Oil on board, 91.5 × 122 cm
Royal College of Art, London

NOTES

1 Auerbach/Tusa interview 2001.
2 *Ibid.*
3 The site may have been the same location as that of cat. 1, or else in the vicinity.
4 Auerbach/Wright interview 2009. One of the paintings he had in mind was Gustave Courbet's *The Painter's Studio*, 1855.
5 See also cat. 1, 2, 4–6.
6 Sylvester 1956, p. 64.
7 Russell 1956.

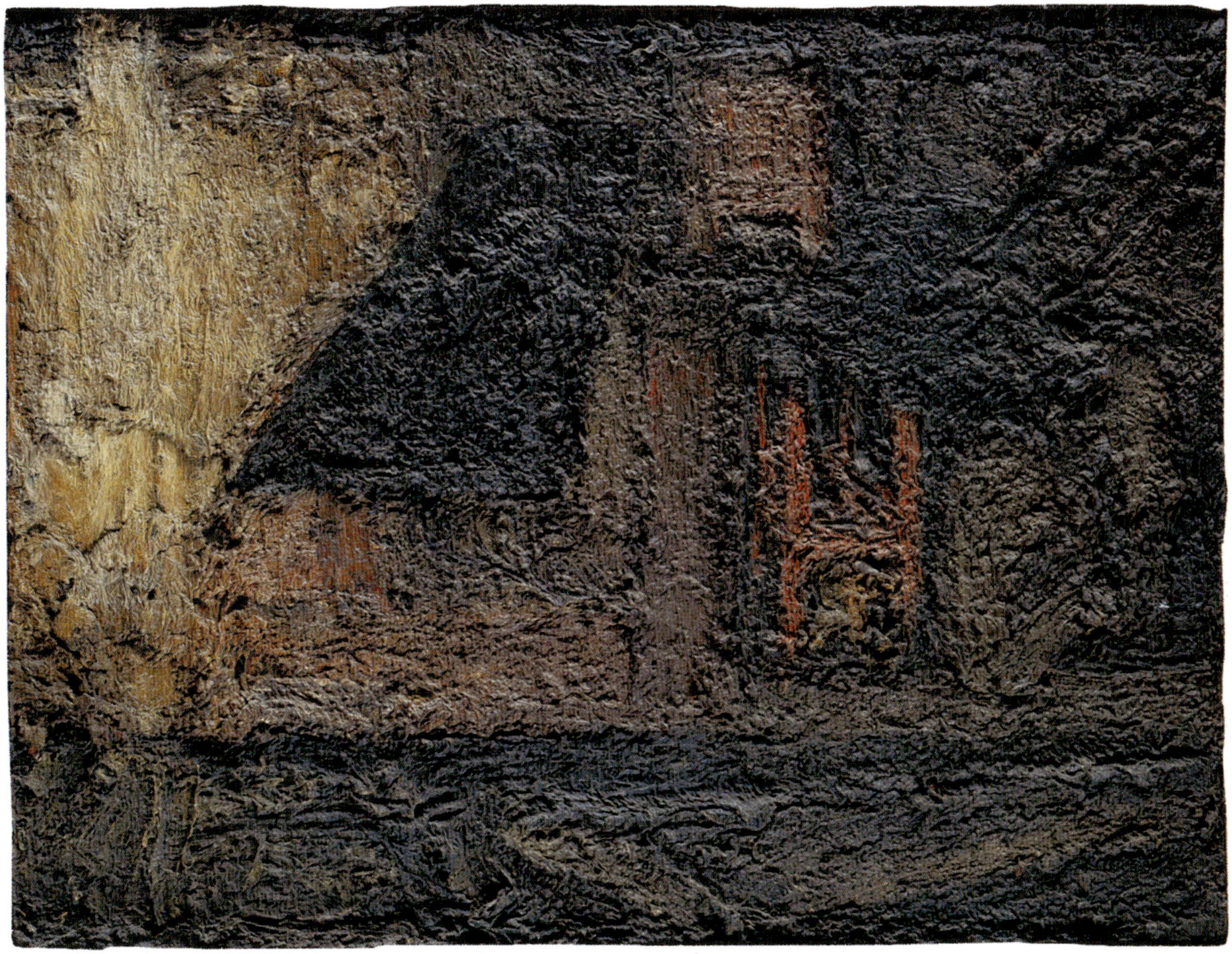

4

Building Site, St Pancras: summer

1954

Oil on board
103 × 128.3 cm
Private collection

SELECTED EXHIBITIONS

'Degree Exhibition', Royal College of Art, London, 1955; *Frank Auerbach*, London, Beaux Arts Gallery, 3–28 January 1956; *Das Junge England*, Europäisches Forum, Alpbach/Tyrol, 19 August–7 September 1960 and Neue Galerie der Stadt Linz, 25 November –end December 1960; *Helen Lessore and the Beaux Arts Gallery*, London, Marlborough Fine Art, February 1968; *Frank Auerbach*, Marlborough-Gerson Gallery, New York, September–October 1969; *Frank Auerbach, Paintings and Drawings 1954–1976*, Marlborough Galerie, Zurich, May 1976; *Frank Auerbach*, Hayward Gallery, London, 4 May–2 July 1978, and Fruit Market Gallery, Edinburgh, 15 July–12 August 1978

This was one of the first building-site paintings completed in Auerbach's Mornington Crescent studio, which he moved into in 1954 – remaining there to this day. The subject was the large construction site of a major residential development, the Regent's Park Estate, which lay a few streets away from his new studio in what was then the London Borough of St Pancras.[1] Following the war, plans were drawn up by St Pancras Borough Council and the London County Council to clear a large swathe of John Nash's famous nineteenth-century terraces in the vicinity of Cumberland Market, which had been badly hit during the Blitz. For this area of some thirty-two acres, the ambition was to create a mixed development of 1,700 flats and houses with provision for shops, pubs and a school. The project's complexity, involving a number of different architects and a tortuous planning process, meant that the development proceeded in a piecemeal way until the early 1960s. Frederick Gibberd, a leading figure of British post-war urban redevelopment and New Town planning, took on the first phase. His series of four- to nine-storey-high blocks of flats were begun in 1951.

When Auerbach visited the site, probably in the summer of 1954, around half of Gibberd's scheme had been built, other phases were well underway and the preparation of enormous foundations for another three blocks of flats eleven storeys high was beginning. These were designed by Davies & Arnold Architects and were for a short time the tallest of their kind in London, though their height would soon be surpassed by even higher towers on the site (fig. 20). An official publication boasted of "the sinking of 880 reinforced concrete piles . . . and while piles were being driven . . . demolition of shops, houses and a public house was forging ahead on other parts of the site."[2] Amidst all this sweeping away of architectural heritage, it was claimed that parts of the development sought to "regain the traditional character of English urban planning" with low-rise terraces providing "ever-changing visual scenery and surprise to the pedestrian". Further high-rise blocks were to be placed "so as to add to the scale and pictorial effect of the layout in the same way as the church spires and towers of the smaller English towns".

The remorselessly muddied tones of Auerbach's painting convey a scene of rawness and desolation that contrasts with the progressive rhetoric of the architects and their ambitions for a brightly Modernist picturesque environment. Of all his building-site paintings this work forges the closest connections between mud and oil paint. The two darkly painted figures of workmen digging out an excavation in the centre of the composition offer a sense of scale, allowing us to make out banks of shored-up earth, which smear into darker cubic forms suggestive of partially completed buildings. Auerbach wanted to use closely related tones to create a particular sense of space. The effect is of space evoked as a totality in which the forms of objects within the composition appear indivisible from the scene's overall environment. What excited him was not the spectacle of modern buildings rising into the sky, but the primal qualities of the earth displaced in the wake of their construction.

NOTES

1 In 1965 St Pancras borough was amalgamated with the boroughs of Holborn and Hampstead to form the borough of Camden.
2 Metropolitan Borough of St Pancras publication on the Regent's Park Redevelopment Area to accompany the visit of H.R.H. the Duke of Edinburgh, 1 December 1955 (unpaginated).

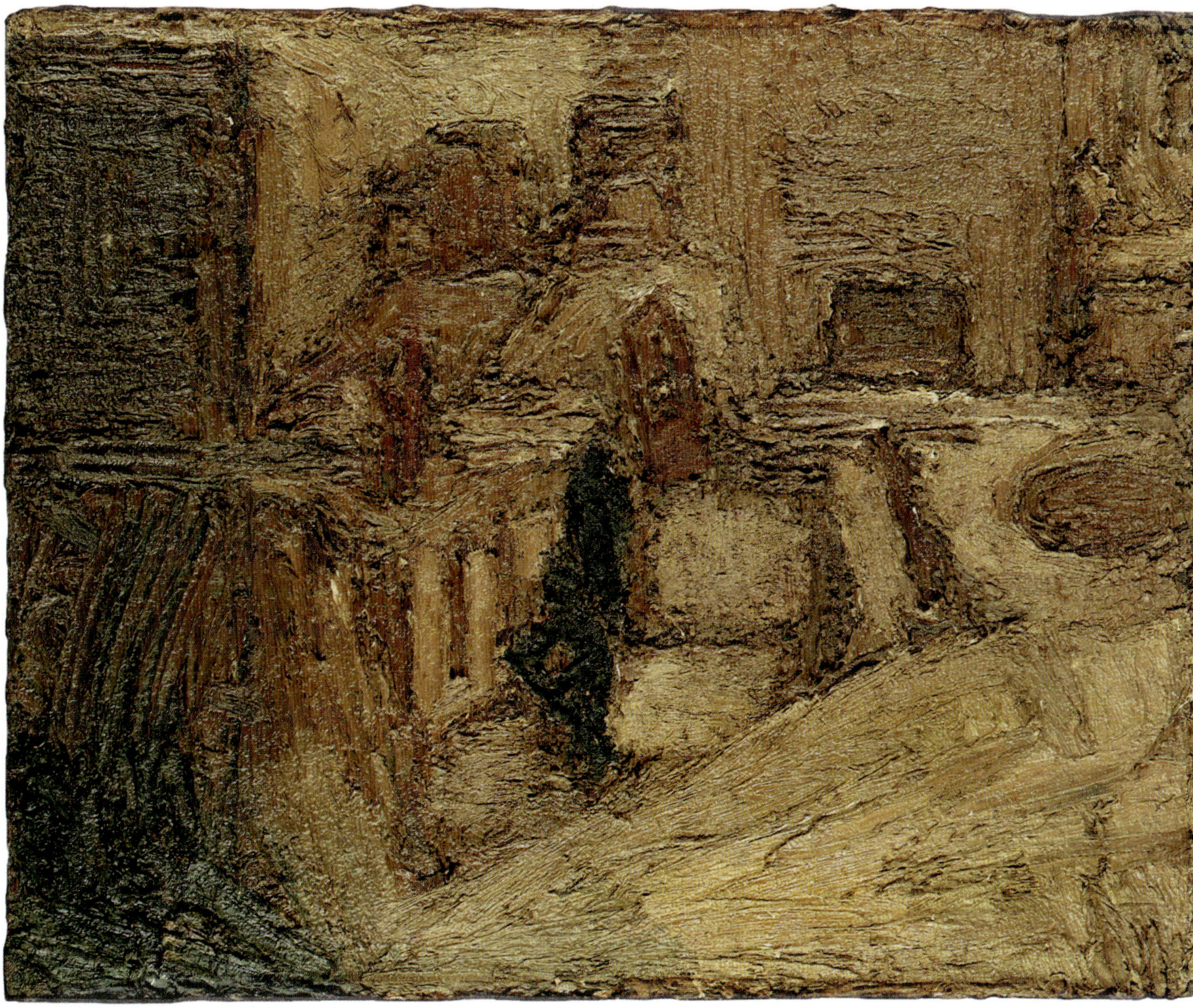

5

Building Site, Portobello Road: winter

1954

Oil on board
104 × 151 cm
Private collection

SELECTED EXHIBITIONS

'Degree Exhibition', Royal College of Art, London, 1955; *Frank Auerbach*, London, Beaux Arts Gallery, London, 3–28 January 1956; *Contemporary British Landscape*, Arts Council, Glynn Vivian Art Gallery, Swansea, 12 November–3 December 1960 (and nationwide tour)

For his first solo exhibition, at the Beaux Arts Gallery in 1956, Auerbach gave a time of year in the titles of his landscape paintings. In addition to simply being informative this may have been intended to underscore the fact that his paintings were fundamentally rooted in direct observation of his subject-matter. Certain reviewers initially found this difficult to discern in his paintings, which seemed to them expressionist or abstract rather than based on *plein-air* sketching.[1] As with all his paintings in the series, the present work would have been the result of perhaps five or six visits to the building site over a short period of time and of around thirty or forty sketches. Auerbach remembered feeling a sense of urgency to gather his required observations in his sketchbooks during these visits: "The thing is about building sites . . . you can't go back because three months later it's an entirely different scene".[2] A selection of these drawings would later be pinned up in the studio and he would begin painting from them. Afterwards Auerbach destroyed almost all of these sketches and so for several of the paintings, including *Portobello Road*, no drawings survive.

The site of this painting is probably the council housing estate of Portobello Court, designed by Armstrong & McManus and built between 1951 and 1953. If so then Auerbach may have made drawings of the site some time during the winter of 1952–53 but not completed the painting until 1954. It was common for him to take many months and sometimes more than a year to finish his paintings. Although they are initially difficult to discern, sustained looking reveals the large figures of workmen dominating the composition. One, rendered in muddy tones, is kneeling in the centre of the image, possibly with a trowel in one hand whilst resting his other arm on his raised knee. Another figure, painted in reds, can be seen standing on the left-hand side perhaps pushing a wheelbarrow, his head and body silhouetted by the whitish forms of distant buildings in the top left-hand corner. Whilst workmen appear in a number of building-site paintings, it is unusual for them to occupy such a prominent position as this. However, their labours were a source of great fascination for Auerbach and offered a sense of continuity with the past: "It was much more handwork and probably closer to the way that people built the pyramids than to the way building sites are conducted now . . . where people working on them are quite often more like machine operatives than manual workers", he recalled recently.[3]

NOTES

1 For example, one reviewer wrote of the building-site paintings, "The careful chaos produced does suggest a vague desolation, but not enough to differentiate between seasons": *The Scotsman*, 23 January 1956.
2 Auerbach/Wright interview 2009
3 *Ibid.*

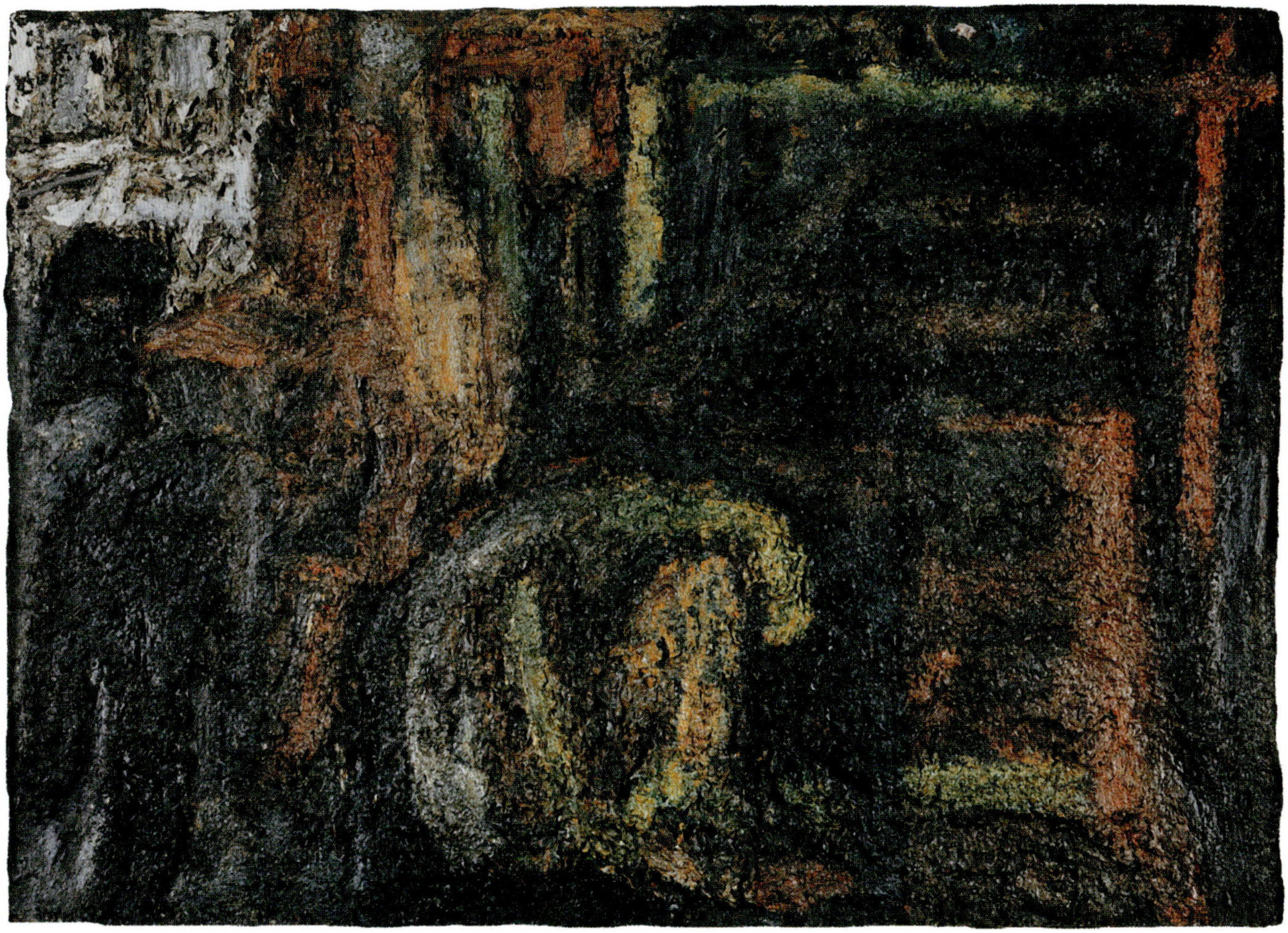

6

Building Site near St Paul's: winter

c. 1955

Oil on board
80.6 × 121 cm
Private collection

SELECTED EXHIBITIONS

'Degree Exhibition', Royal College of Art, London, 1955; *Frank Auerbach*, London, Beaux Arts Gallery, London, 3–28 January 1956

St Paul's Cathedral was the enduring symbol of London's resilience during the Blitz. Herbert Mason's photograph of its iconic dome emerging steadfast and unscathed amidst the flames and smoke of a German bombing raid became a defining image of the war. Bomb-damage maps (see fig. 16) reveal just how miraculous its survival was, with almost the entire area around it being razed to the ground in successive waves of air raids between 1940 and 1941. The first of these bombed areas next to St Paul's to be redeveloped after the War was the land at the east end of the cathedral, bordered by Watling Street and New Change (see fig. 22). It was here, where two adjacent construction sites were being developed for the large office blocks of Gateway House and New Change House, that Auerbach made sketches for the present painting. As the first buildings to go up on this hallowed ground around St Paul's both blocks attracted considerable interest. Gateway House was completed first, in 1956, but its designs, by Trehearne & Norman and Preston & Partners, had already drawn considerable criticism in some quarters for their historicising style, considered retrograde. New Change House, by Victor Heal & Smith, started in 1953 and finally completed in 1960, confirmed the view that the buildings' neo-Georgian façades represented a loss of modern architectural nerve through excessive reverence for St Paul's. "Respect for neighbours," argued J. M. Richards, "does not lie in stylistic devices, but in proper control of massing, scale and silhouette."[1]

A sketch relating to the painting (fig. 37) shows the emerging steel frame of New Change House at a slightly earlier phase in the building's development than that shown in a photograph from 1955 (fig. 36). In the drawing Auerbach was attracted by the contrast between the crossed arms

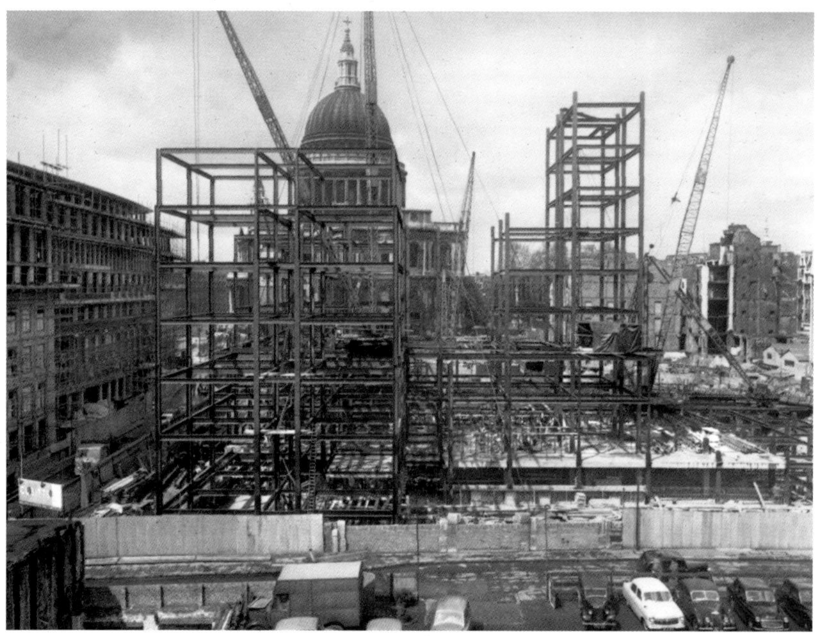

Fig. 36
New Change House under construction, 1955
Photograph: Hulton Archive/Getty Images

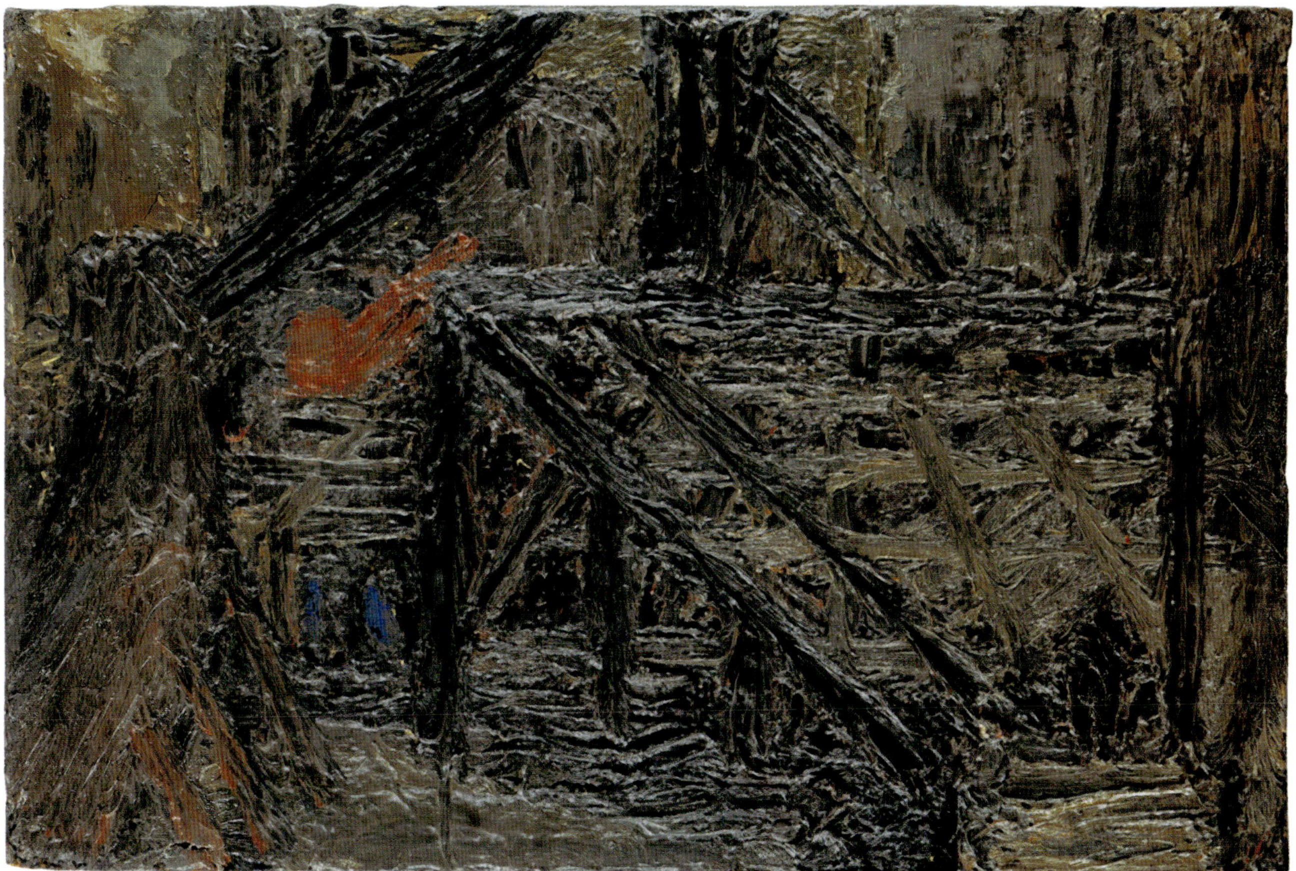

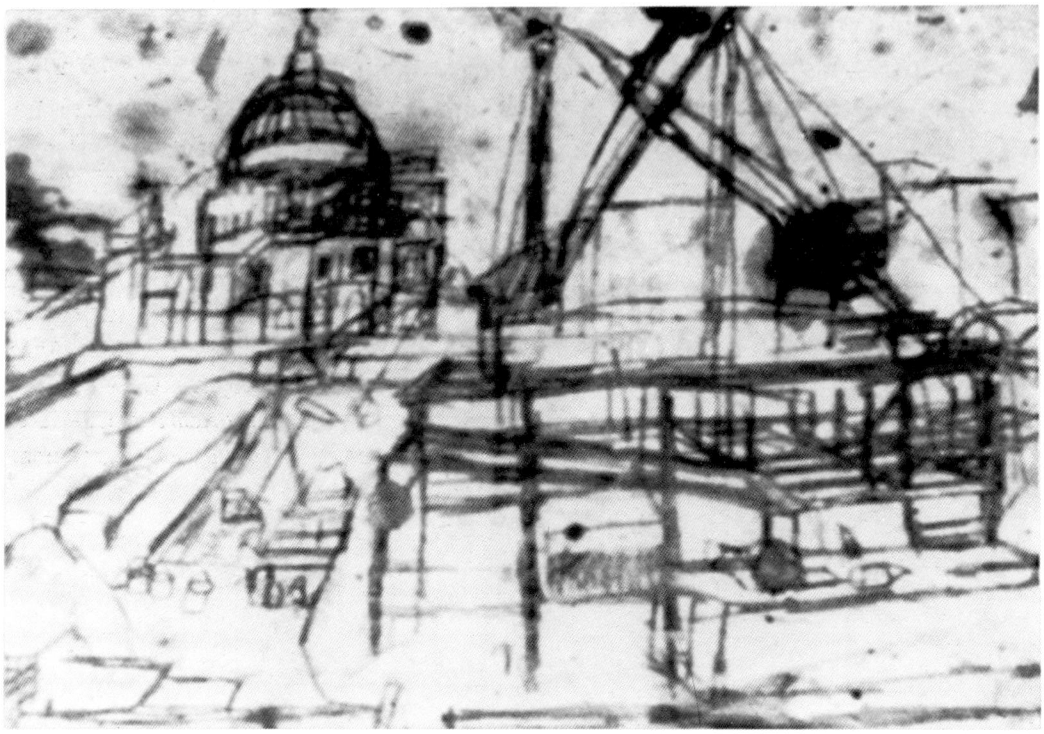

Fig. 37
Frank Auerbach, *Sketch for Building Site near St Paul's: winter*
c. 1955
Pencil and ink on paper, 27 × 38 cm
Private collection
(Archival photograph)

of the cranes lifting steel beams into place and the monumental form of the Cathedral itself. However, in the painting he adjusted his angle of view considerably and St Paul's is almost completely obscured by the dense black grid of cranes and steel, its dome and apse just glimpsed towards the top left of the composition. There is something almost deliberately anti-picturesque about Auerbach's decision to smother the cathedral in this impenetrable web of construction, although his concerns would undoubtedly have been formal ones. His close friend Leon Kossoff, who also produced a large body of building-site works, also painted the New Change site at this time (fig. 31). However, Kossoff accorded St Paul's a very different status, picturing it rising majestically above the chaos below, as David Bomberg had done in his wartime painting, *Evening in the City* (fig. 32). Auerbach's apparent indifference to the symbolic potency of his subject expresses itself forcefully, and contrasts with the prevailing post-war veneration of St Paul's as the city's emblem of historical continuity and survival. It also contrasts starkly with the neo-Romantic renderings of Britain's bombed churches and other historical sites by John Piper, which were perhaps the most famous modern paintings of these wartime subjects (see fig. 4). And yet, far from distancing the viewer from the subject, Auerbach's unconventional composition and his insistence on the formal and material qualities of his painting intensify and renew the viewer's engagement with the historically charged site.

NOTE

1 J.M. Richards, 'Rebuilding the City', *Architectual Review*, August 1954, p. 118.

7

Shell Building Site: Workmen under Hungerford Bridge

c. 1958–61

Oil on board
111.7 × 139.7 cm
Private collection

7 (i)

Study for Shell Building Site: Workmen under Hungerford Bridge

c. 1958

Pencil on paper
21.5 × 28 cm
Private collection

SELECTED EXHIBITIONS

(*Shell Building Site: Workmen under Hungerford Bridge*):
Frank Auerbach, London, Beaux Arts Gallery, November–December, 1959

The Shell Building site was one of the largest and most dramatic areas of post-war rebuilding in central London: Auerbach described it as being like "the Grand Canyon" (see cat. 8).[1] Developed on the site of the 1951 Festival of Britain as one of the central offices of the Royal Dutch Shell Company, the building works stretched along the South Bank from Waterloo Bridge to County Hall. Designs for the project by Sir Howard Robertson and Maynard Smith were first exhibited in 1956 and work commenced the following year. They proposed two office complexes, one L-shaped block to be sited to the east of Hungerford Bridge and the other a much larger U-shaped block to the west of the bridge, which would boast a 26-storey tower – London's first skyscraper. The tower would be the first office block to exceed the height of the Palace of Westminster's Victoria Tower and it came close to the height of the spire of St Paul's Cathedral. During its construction the building drew criticism from various quarters. Some commentators feared the high-rise precedent that it set and the impending Americanisation of London's architecture.[2] For others it was not the issue of height *per se* but the design and proportions of the complex that was at fault, Reyner Banham describing it before construction even began as "South Bank's Vertical Failure". Completed in 1962, the Shell Building was then the largest office complex in Europe and epitomised the office building boom of the 1950s and 1960s that was changing the face of London forever.

Auerbach visited the site on many occasions during the early stages of its development, making sketches of the whole area from different vantage points. Only a few of these survive, including the study for this work (see also cat. 8 (ii) and 8 (iii) and 9 (i)). The drawings led to three paintings of the site, which were exhibited together at his 1959 Beaux Arts Gallery solo exhibition (see cat. 8 and 9). For the present work he made the nineteenth-century railway arches of Hungerford Bridge his principal subject. Railway arches held particular interest for Auerbach, who was also painting his two versions of *Railway Arches, Bethnal Green* at this time (see fig. 7). It is likely that Auerbach positioned himself on the eastern side of the bridge, behind the Royal Festival Hall, where a large area of steel-reinforced concrete foundations was under construction; these appear in the foreground of Auerbach's composition and can be seen in their full extent in a 1958 photograph (fig. 38).

When first exhibited at his 1959 Beaux Arts Gallery show, the painting was rendered in earth tones. In this form it was singled out for praise by the *Times* reviewer, who admired the "almost purple bloom" that the artist had created with brown pigments, "which achieves impressive effects not only of depth and space and light . . . but of colour".[3] However, Auerbach was dissatisfied with it. "It just seemed timid," he recalled, "as though it was done from the shoulder."[4] He took it back from the gallery and it remained for a long period in his studio, while he gathered up the courage to repaint it. "I must have had some brandy because I remember drinking and repainting the whole thing in black and white from top to bottom . . .

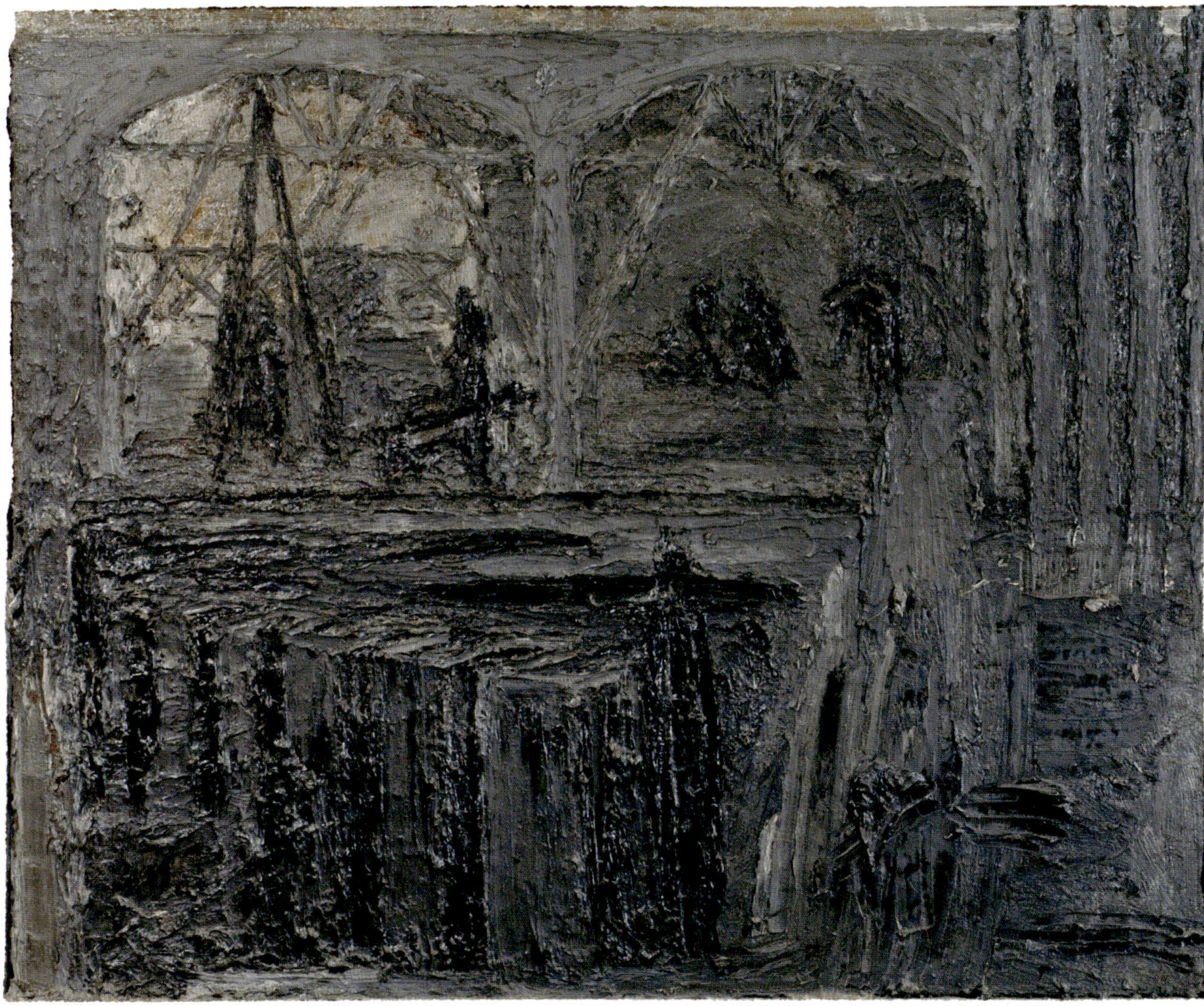

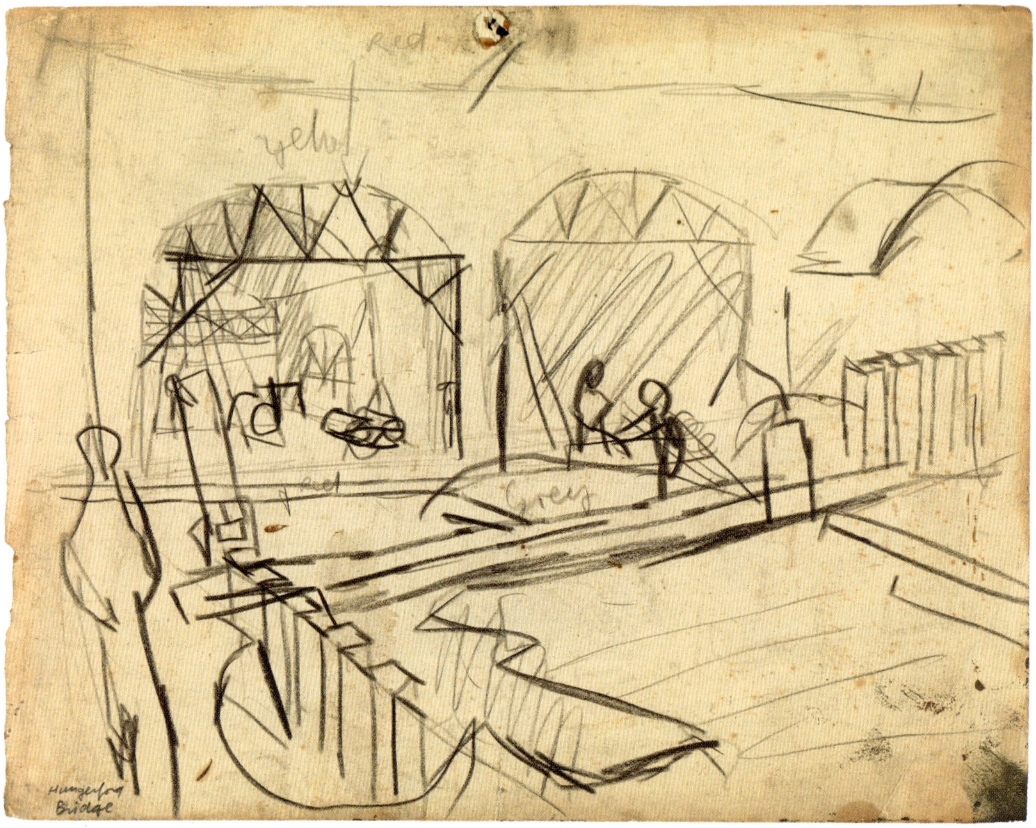

7 (i)

Fig. 38
The Shell Building construction works on the east site behind the Royal Festival Hall, with Hungerford Bridge running across the centre, 1958
Photograph: Hulton Archive/Getty Images

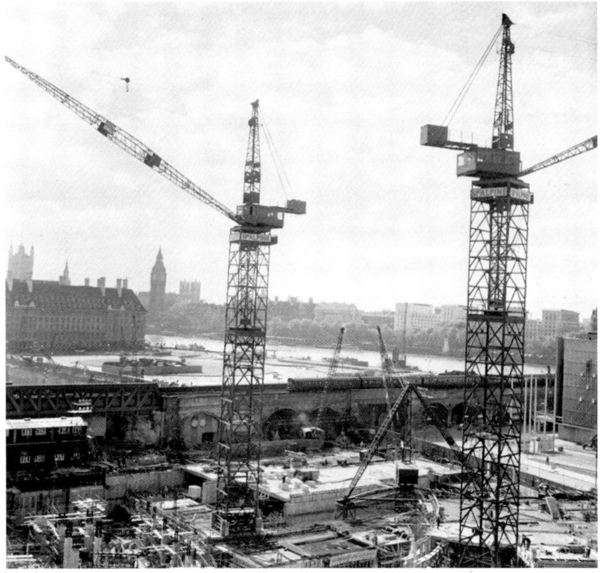

in a day!"[5] The spontaneous reworking of paintings after months of laborious endeavour was not unprecedented in Auerbach's work (see cat. 1 and 2); however, the speed and decisiveness with which he transformed *Workmen under Hungerford Bridge* was highly uncharacteristic. And yet the resulting image, shrouded in black tones, does not obviously express the speed of its completion. Rather, it offers a sepulchral vision in which forms gradually emerge through the darkness as one's eyes become accustomed to Auerbach's light. Texture also plays a particularly important role. Areas of the surface have dried to an almost chalky finish – reminiscent of his contemporary charcoal drawings of heads – but where he has broken through this dry skin to pick out salient details the paint appears glossy, like the oozing sap of tree bark. The effect is to distance the viewer from the immediacy of the subject-matter and to engage one in a slow process of contemplation and reconnection in which the visual and the haptic conjoin. David Sylvester described a similar experience in relation to Auerbach's figure work, as "like that of running our fingertips over the contours of a head near us in the dark, reassured by its presence, disturbed by its otherness, doubting what it is and then whether it is."[6]

Although Auerbach was driven by formal concerns, the work's sombre palette and charred-looking forms are suggestive of themes of life and death. Set in the context of the site itself, the work may connect profoundly with the lives and deaths of this particular area of the South Bank. The site had been destroyed and rebuilt in its near entirety no less than three times in little over a decade: its nineteenth-century housing and docks were bombed during the Blitz of 1940–41, then bulldozed and redeveloped as the Festival of Britain in 1951, and then levelled once again shortly afterwards for the construction of the Shell Building. Indeed the only remnant of its earlier history was Hungerford Bridge itself.

NOTES

1 Auerbach/Wright interview 2009.
2 Robertson had been one of the team of architects who worked on the United Nations Building in New York.
3 Thompson 1959, p. 14.
4 Auerbach/Wright interview 2009.
5 *Ibid.* The painting may have been exhibited in its final form in the 1961 Beaux Art Gallery mixed summer show, although there was no published catalogue. Beaux Art Gallery records date the painting to 1961: Marlborough Gallery Archives.
6 Sylvester 1956, p. 64.

8

Shell Building Site: from the Festival Hall

1959

Oil on board
124.5 × 152.4 cm
Gray Art Gallery and Museum, Hartlepool

8 (i)

Shell Building Site: from the Festival Hall (Oil Sketch)

1959

Oil on board
37 × 42.2 cm
Private collection

SELECTED EXHIBITIONS

(*Shell Building Site: from the Festival Hall*): *Frank Auerbach*, London, Beaux Arts Gallery, November–December, 1959; *Cityvisions: Urban Landscape Painting in Britain 1951–82*, Bolton Museum and Art Gallery, 12 March–16 April 1983
(*Shell Building Site: from the Festival Hall* (*Oil Sketch*)): *Frank Auerbach*, London, Beaux Arts Gallery, November–December, 1959

"Fantastic! It was superb! It was a vast building site and it looked absolutely superb. It was like the Grand Canyon. I remember going there ... it was almost a gift, the thing was so superb in itself that you could have taken it and put it in a museum being what it is, such a marvellous thing."[1]

For Auerbach, discovering the mountainous landscape of the Shell Building construction site was the culmination of his search for the most dramatic and awe-inspiring of London's building sites (see cat. 7). The thrill of the spectacle of deep excavations with narrow ridges and sheer drops was mixed with fear: "I've no head for heights .... I remember going to places where I was scared to stand, and drawing. I remember that on these planks people wheeled wheelbarrows totally confidently, I would sit down and edge my way along them in order to do my drawings."[2] This potent mix of excitement and fear – a defining characteristic of the sublime landscape tradition that Auerbach greatly admired, especially in the work of Turner – led to two paintings of the massive earthworks on the western part of the site (see also cat. 9).

For the present work Auerbach took a vantage point at the Royal Festival Hall end of the site (see fig. 40) looking across and down into one of the deep excavations which would eventually form the foundations and basement levels of the Shell Building's 26-storey skyscraper tower (fig. 39; see also figs. 23 and 24). The two surviving pencil drawings and an oil sketch give a rare insight into Auerbach's working practice. The drawings demonstrate how he moved between recording closely observed details of the scene, including cranes, ladders, workmen and

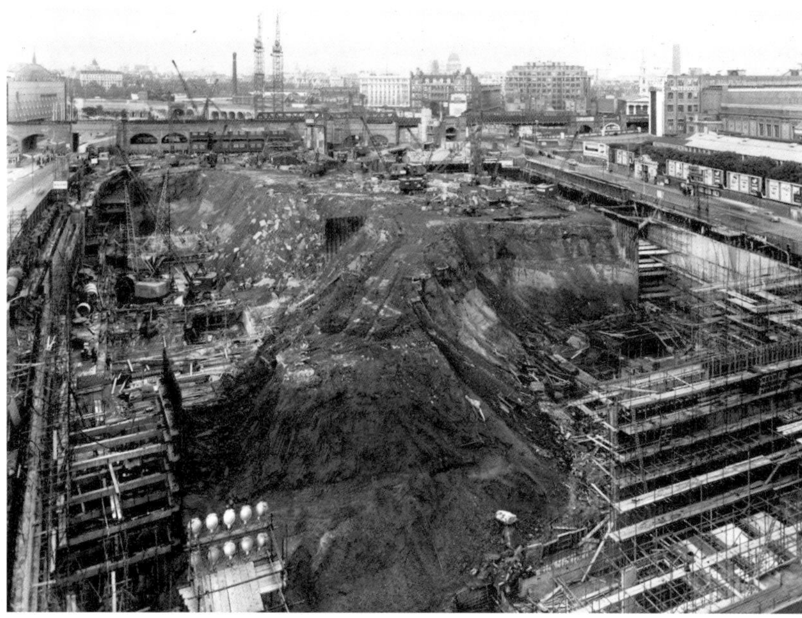

Fig. 39
The Shell Building construction works showing the excavations for the foundations of the tower on the west area of the site, *c.* 1958–59
Photograph: Sir Robert McAlpine Ltd

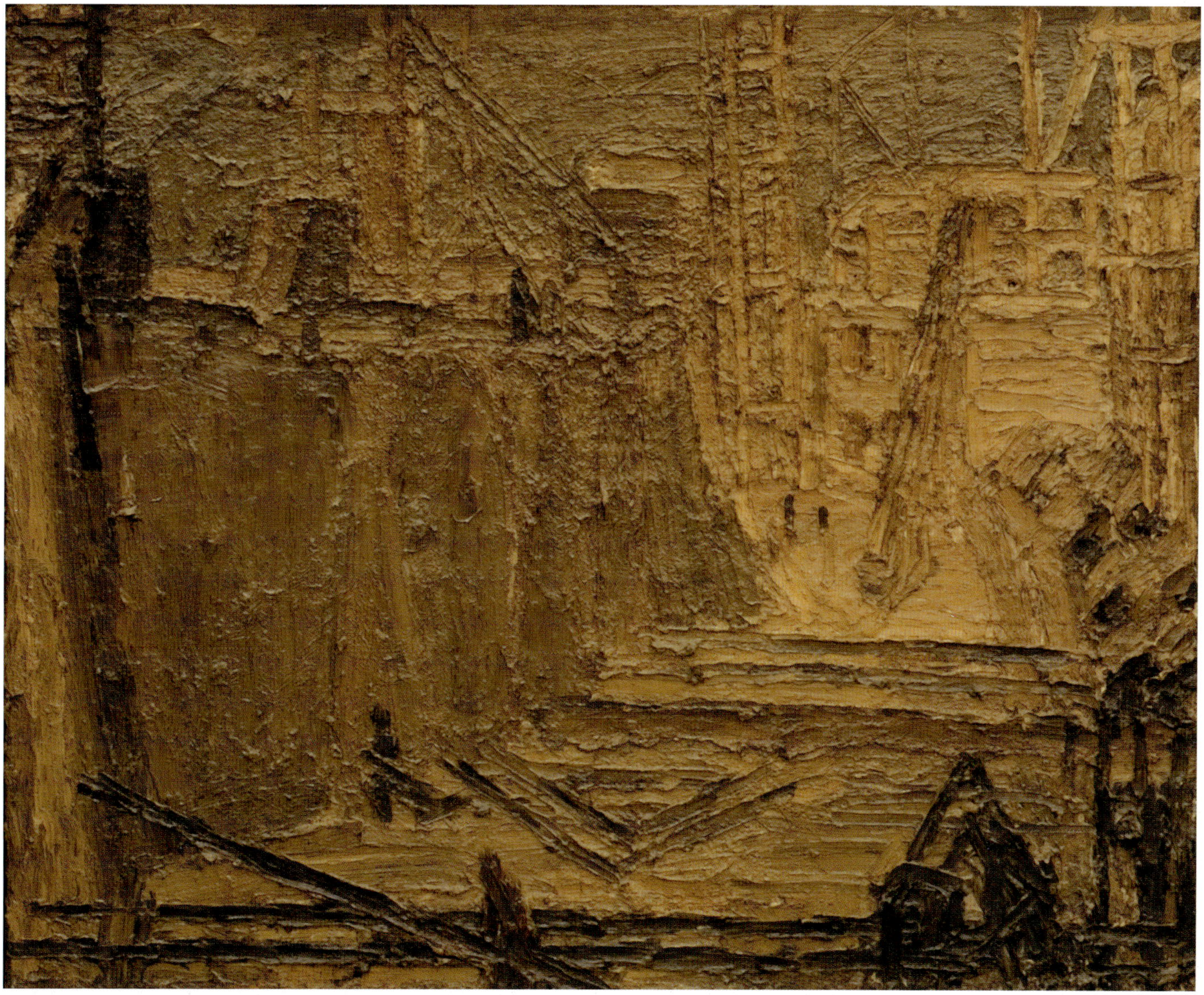

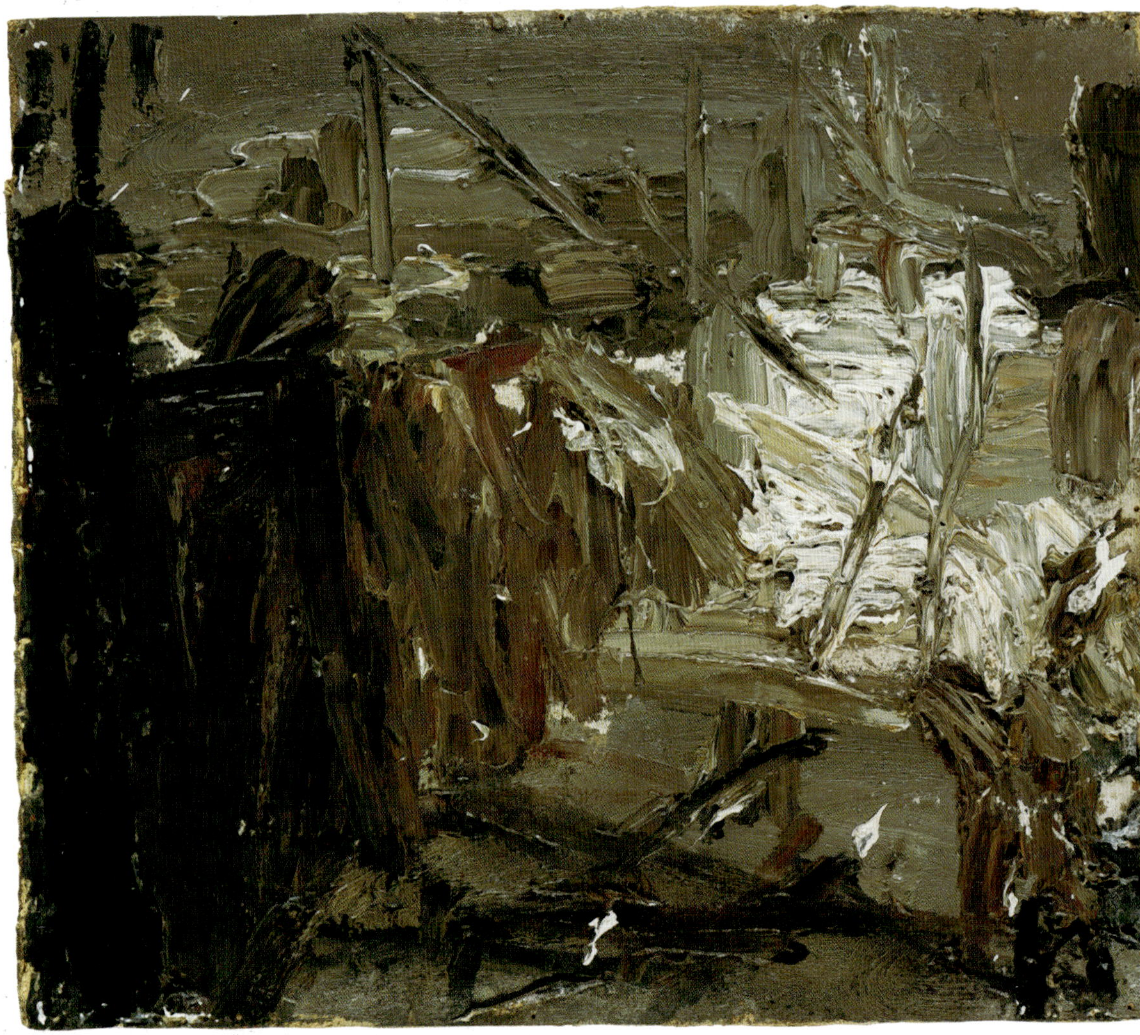

8 (i)

8 (ii)

Study for Shell Building Site: from the Festival Hall

c. 1958–59

Pencil on paper
17.8 × 22.5 cm
Private collection

8 (iii)

Study for Shell Building Site: from the Festival Hall

c. 1958–59

Pencil on paper
19 × 22 cm
Private collection

other equipment (cat. 8 (iii)), to rapidly scribbling down the basic forms of the composition (cat. 8 (ii)). Already in this most cursory of sketches he establishes a clear sense of the density and lighting of different areas of the scene that will be carried through to the finished painting. Much of the detail captured in the other drawing is also clearly retained in the painting. The pin holes in both sheets remind us that the two sketches would have been put up in the studio together and, along with others, would have provided constant reference points for Auerbach as he worked on the painting.

The oil sketch (cat. 8 (i)) would not have been worked up from drawings as an intermediary stage before he commenced the large painting, as was conventional artistic practice. Rather, it was typical for Auerbach to make an oil sketch when work on the principal painting had stalled or become difficult. The small scale of the sketch gave him the freedom to explore and resolve a problematic area, enabling him to return afresh to the large work. The oil sketch, like the rapidly scribbled drawing, establishes the basic structure of the composition through the balance of light and dark tones. Sometimes Auerbach would deem an oil sketch successful in its own right and he would complete it as an independent work, as in this case. Five building-site oil-sketches were finished in this way, and all were exhibited at Auerbach's various Beaux Arts Gallery exhibitions along with the principal paintings to which they related (see cat. 11 (i–iii) and fig. 42).

NOTES

1 Auerbach/Wright interview 2009.
2 *Ibid.*

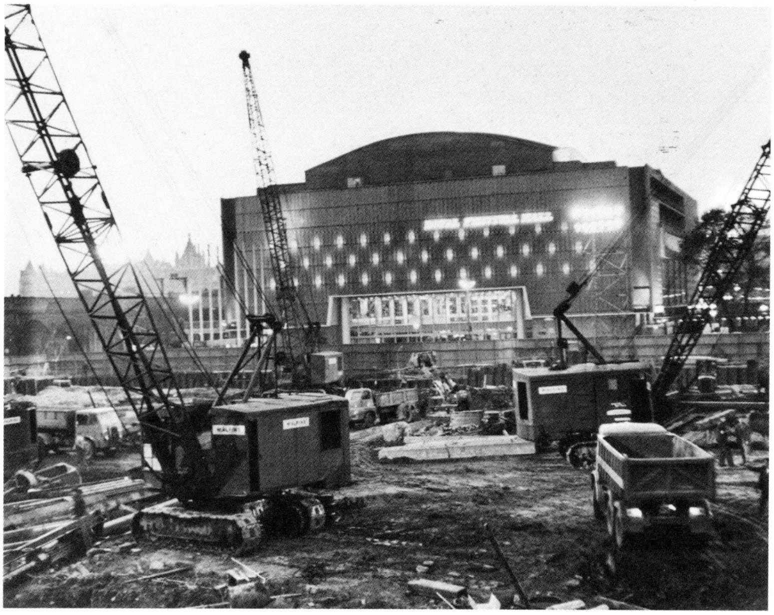

Fig. 40
The Shell Building construction works behind the Royal Festival Hall on the east area of the site, *c.* 1958–59
Photograph: Mary Evans Picture Library

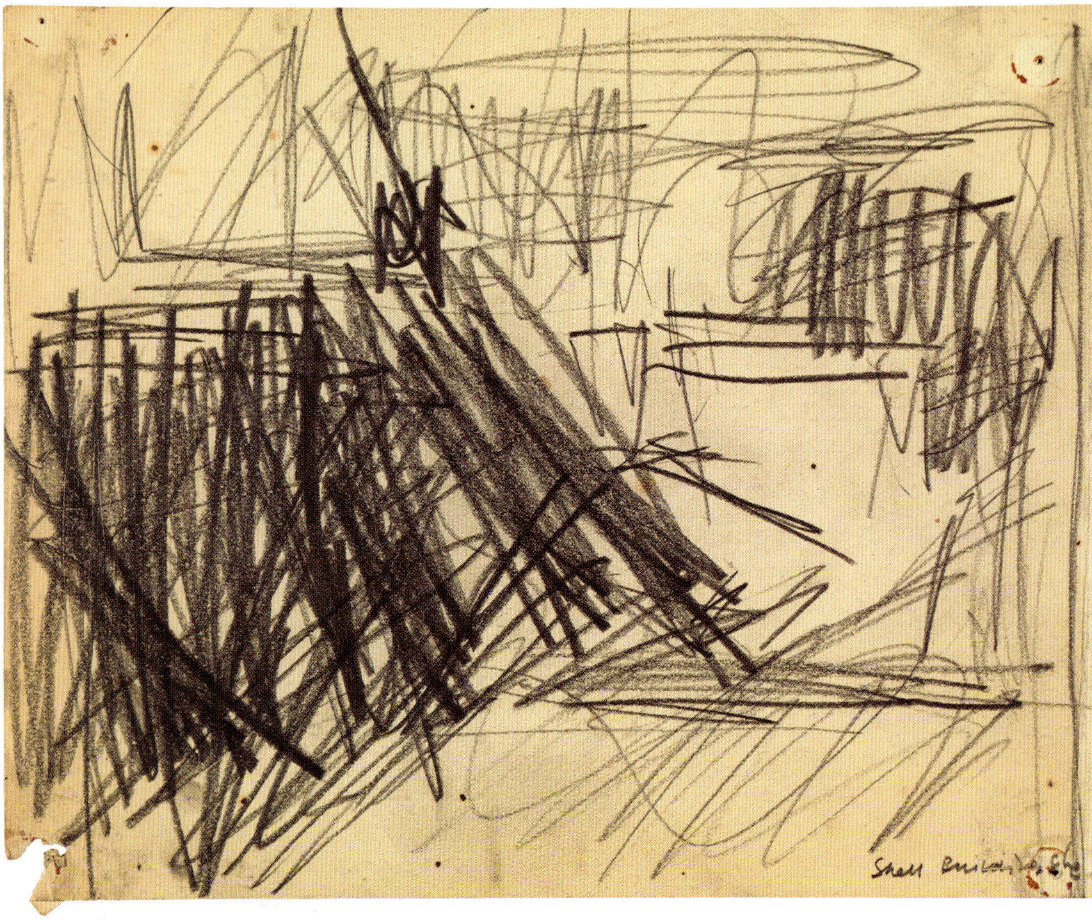

8 (ii)

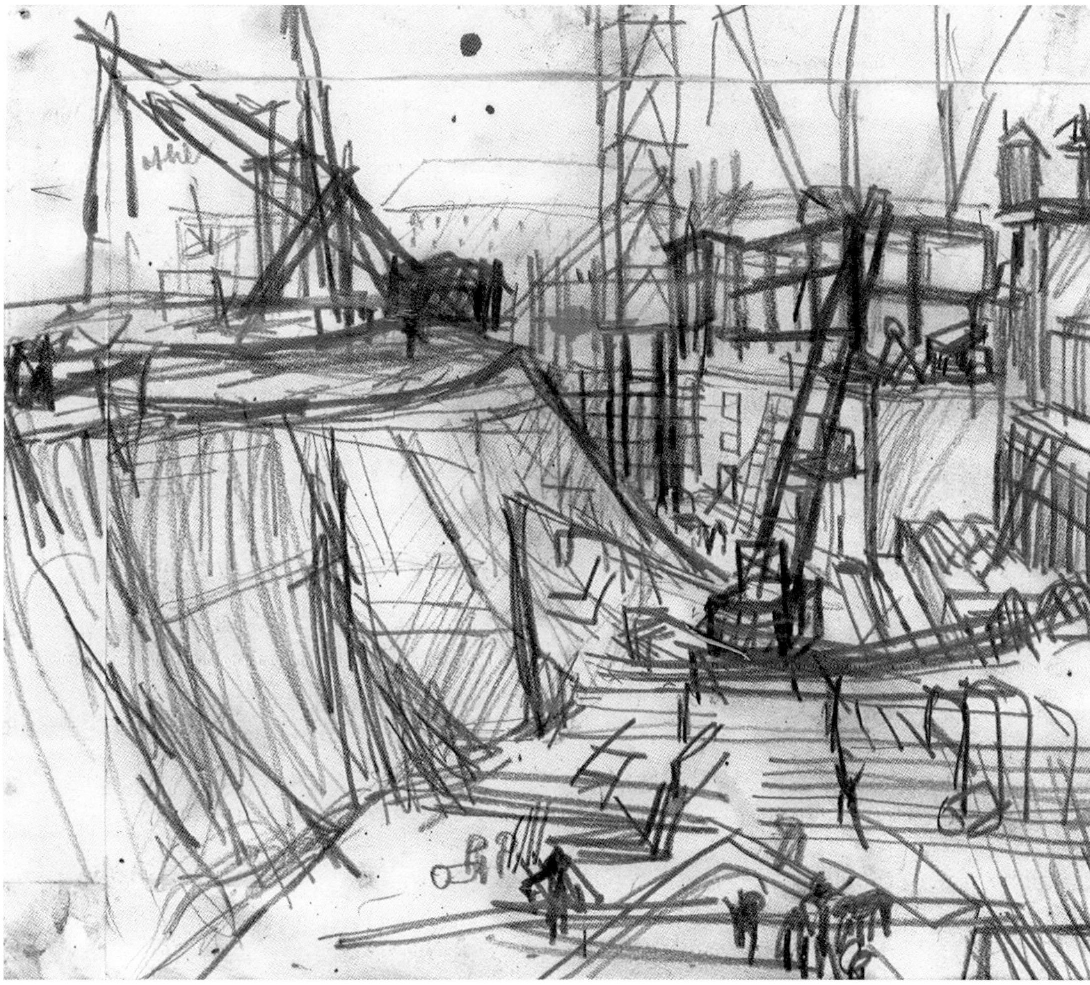

8 (iii)

9

Shell Building Site: from the Thames

1959

Oil on board
152.4 × 121.9 cm
Museo Thyssen-Bornemisza, Madrid

9 (i)

Study for Shell Building Site: from the Thames

1959

Pencil on paper
20.9 × 17.7 cm
Private collection

SELECTED EXHIBITIONS

(*Shell Building Site: from the Thames*): *Frank Auerbach*, Beaux Arts Gallery, London, November–December 1959; *Frank Auerbach*, Hayward Gallery, London, 4 May–2 July 1978, and Fruit Market Gallery, Edinburgh, 15 July–12 August 1978; *The Hidden Trace: Jewish Paths through Modernity*, Felix-Nussbaum-Haus, Osnabrück, 7 December 2008–19 April 2009

This is one of Auerbach's most celebrated building-site paintings. Its dramatic rendering of a crane's boom dropping its cable into the bowels of the earth, which seem to radiate light from within, is a powerful evocation of the awesome spectacle of the enormous Shell building site (see also cat. 7 and 8). The surviving pencil sketch for the composition reveals Auerbach's instinctive urge to form a triangle out of the lines of the crane's arm and cable by adding a third line running diagonally down the side of the earth bank. This triangular shape becomes a defining element in the finished painting and is extended to create a further symmetrical triangle, formed in part by another crane arm to the right. Auerbach's stated desire to find "a secret geometry"[1] from within his images is expressed particularly vividly through this kite-shaped network of bisecting lines. They find echoes in different forms throughout the painting, for example in the saw-toothed roofline of Waterloo Station in the background and in further cranes and structures on the right-hand edge.

The painting is also a vivid expression of Auerbach's profound engagement with the work of Rembrandt. As well as regularly sketching the streets and building sites of London, he was a frequent visitor to the National Gallery, where he dedicated himself to sketching after the Old Masters. "I thought about Rembrandt all those years," he recalled; "one's head was full of all the paintings one admired, but of all the Old Master paintings it was Rembrandt I was most moved by."[2] Rembrandt's small painting *The Lamentation* (fig. 41) captivated Auerbach especially, and it was around this time that he set about painting a large-scale study after it (fig. 43), in which vectoring lines, suggested to him by Rembrandt's

Fig. 41
Rembrandt Harmenszoon van Rijn,
The Lamentation over the Dead Christ
(also known as *The Deposition*)
c. 1635
Oil on paper and pieces of canvas mounted on oak panel, 31.9 × 26.7 cm
The National Gallery, London

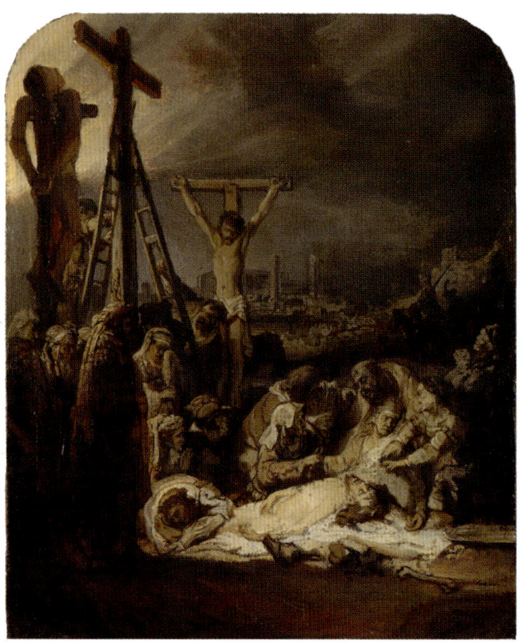

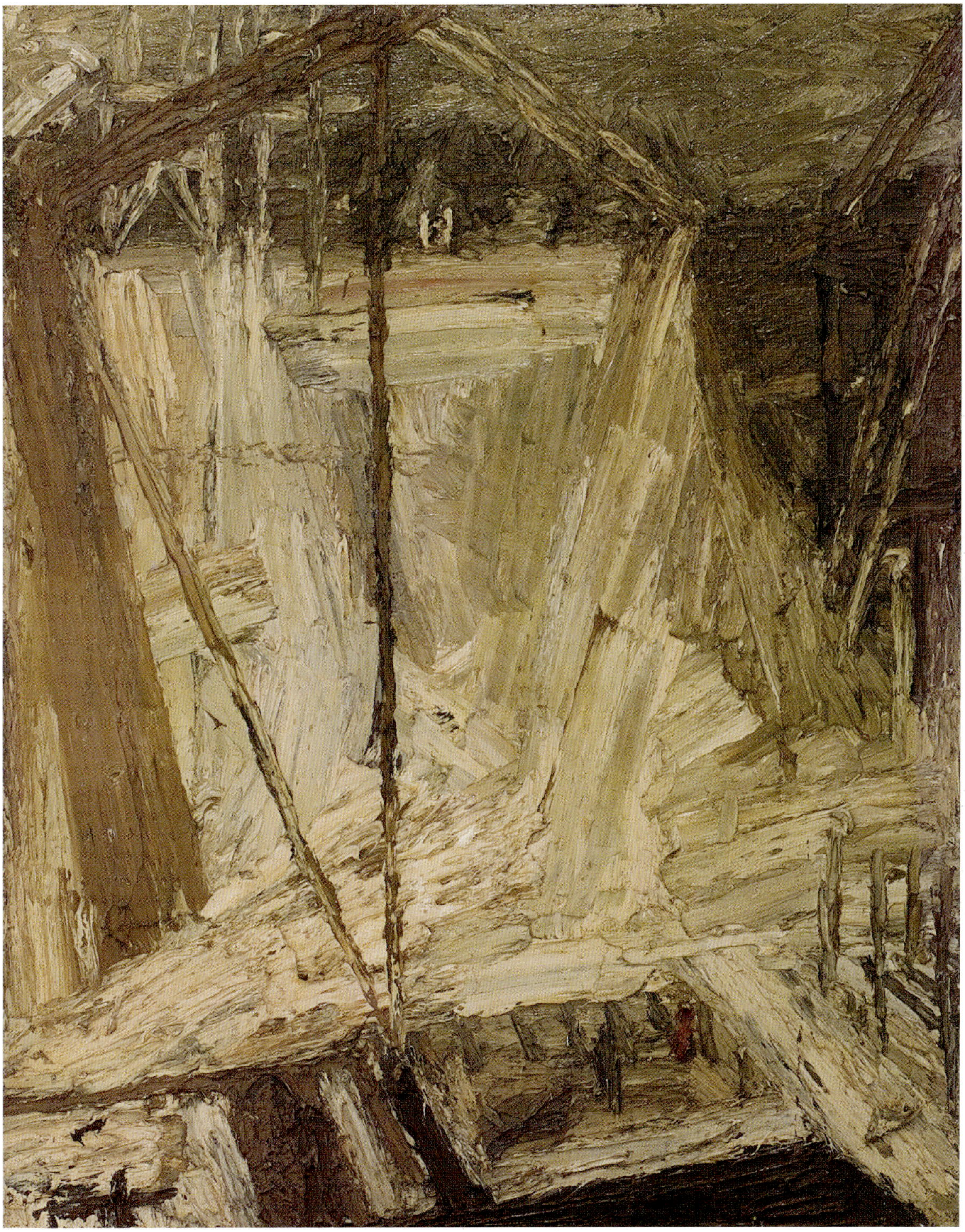

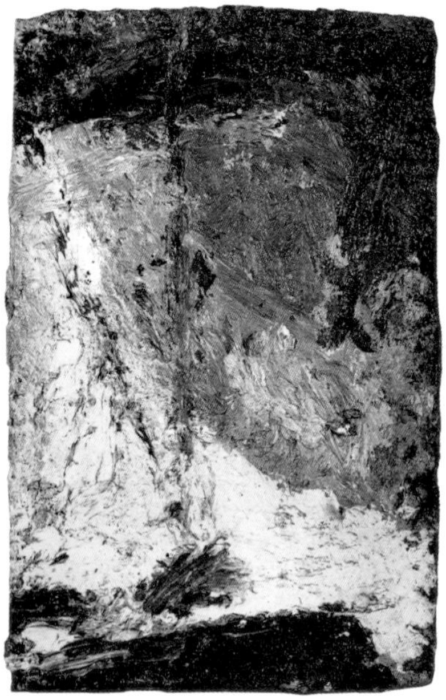

Fig. 42
Frank Auerbach, *Shell Building Site: from the Thames (Oil Sketch)*
1959
Oil on board, 30.5 × 20 cm
Private collection
(colour photograph unavailable)

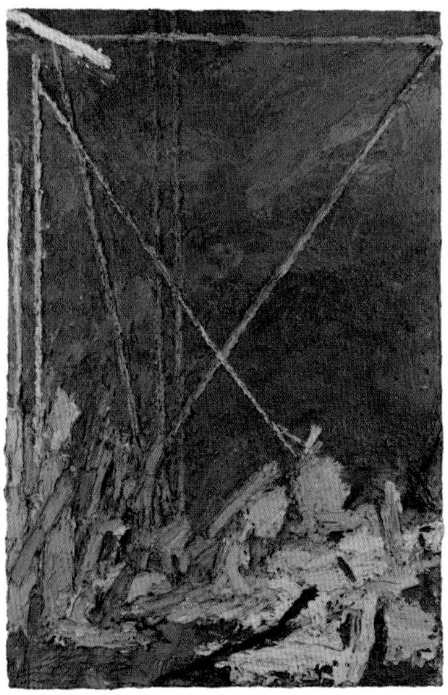

Fig. 43
Frank Auerbach, *Study after The Deposition by Rembrandt II*
1961
Oil on board, 180.3 × 121.9 cm
Private collection

original composition of crucifixes and ladders, extend across the image. Both *The Lamentation* and his transposition of it have clear connections with *Shell Building Site: from the Thames*. During the lengthy course of Auerbach's reworking of the painting, the form of the crawler crane positioned on the top edge of the bank was transformed by the strengthening of a brownish band in the cut of earth beneath it, the whole now recalling a crucifix. Moreover, the dramatic use of light radiating from the base of the excavation in Auerbach's painting echoes the way in which light mystically emanates from the dead body of Christ, illuminating Rembrandt's composition. The oil sketch for Auerbach's work (fig. 42) appears to have played an important role in developing the painting's 'Rembrandtesque' use of light; it is also of a similar scale to the *Lamentation*.

Equally immersed in Rembrandt's *Lamentation* at the National Gallery and in the chasms of earth and forest of cranes at the Shell Building site, Auerbach produced a painting which re-imagined both, imbuing the work with themes of death and resurrection. These themes lie at the heart of all Auerbach's building-site paintings, both in terms of his destructive process of their creation, constantly working and reworking the image, and as a response to the city itself, rebuilding from the debris of war. The relationship between the scene of Christ's deposition and the lamentation over him in the foreground of Rembrandt's painting and the ruined city in the background must have resonated particularly strongly with Auerbach at this time.

The painting also marks the beginning of a more dynamic and gestural character to Auerbach's work, though it involved as much working and reworking as his earlier paintings. Whilst clearly part of Auerbach's engagement with an Old Master tradition, *Shell Building Site: from the Thames* offers a distinctive contribution to the concerns of international contemporary painting and prompts particular comparison to American Abstract Expressionism, which had been shown in London for the first time in 1956.[3]

NOTES

1 Auerbach/Wright interview 2009; see also cat. 2.
2 Auerbach/Wright interview 2009.
3 *Modern Art in the United States: A Selection from the Collections of the Museum of Modern Art, New York*, The Tate Gallery, London, 5 January–12 February 1956.

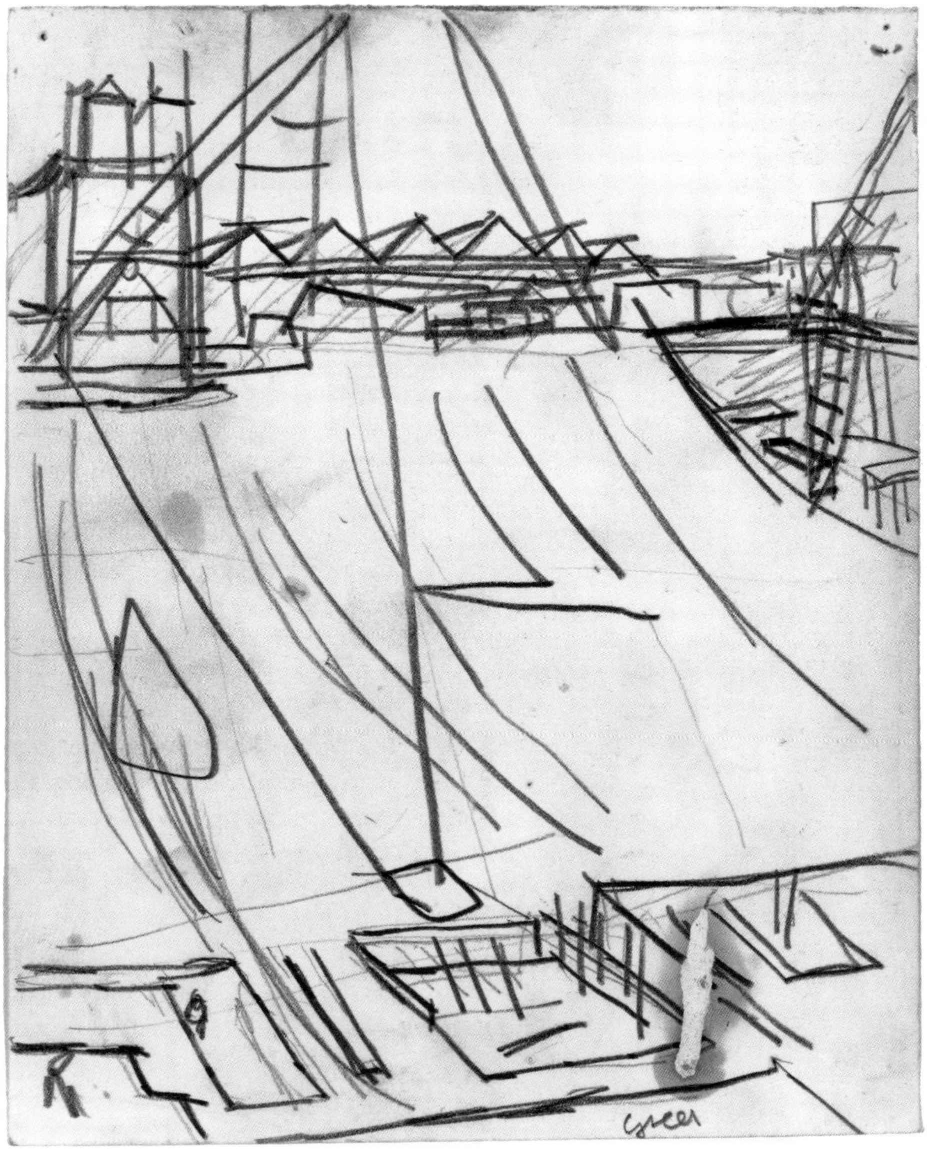

9 (i)

10

Victoria Street Building Site

1959

Oil on board
112.4 × 150.5 cm
Ferens Art Gallery, Hull

SELECTED EXHIBITIONS

Frank Auerbach, Beaux Arts Gallery, London, November–December 1959; *Frank Auerbach*, London, Beaux Arts Gallery, April–May, 1961 (not listed in published catalogue but added as a hand-written amendment to the Beaux Art Gallery's copy); *Recent Figurative Painting*, University of Nottingham, 13 January–7 February 1960; *Transition: The London Art Scene in the Fifties*, Barbican Galleries, London, 31 January–14 April 2002

This is one of the most quickly produced large paintings that Auerbach has ever completed. He made drawings of the site and started work on the painting soon afterwards, completing it in around six or seven weeks. By comparison, other works typically took up to a year or more to finish. Accordingly, the paint surface is not quite as encrusted as others in the series, although the impasto is still thick and heavily worked by conventional standards. Whereas today Auerbach's practice is to scrape down his paintings before reworking them, during this period he did not have the courage to destroy his previous day's image in this way and so the longer he painted the thicker a work became. "I didn't want them to be thick, they just came out thick," he explained. "Every time I started another painting I thought, 'Well this will be a thin one', and it took a long time before one of them was .... It has taken me sixty years to get a reasonable amount of paint on the canvas."[1] The large quantities of paint required to supply his way of working meant that for much of the 1950s his palette was largely restricted to cheap earth tones rather than expensive coloured pigments, which he was only able to start buying regularly from around 1962, when a Beaux Arts Gallery contract for £1500 a year gave him increased financial means (see cat. 13). The parallel between the churned muddiness of London's building sites and Auerbach's thick, earth-toned paintings of them was the result of economic and artistic necessity rather than preconceived design.

This site on Victoria Street was probably that of a large office block designed by Ronald Fielding which would house the Board of Trade (later Department of Trade and Industry), not far from Westminster Abbey.[2] Some commentators criticised the project on the grounds that the area of development was not a bomb site and required the demolition of the nineteenth-century Westminster Chambers building. For *Architects' Journal* the project was further proof of "the rampant speculation in office building", which, it claimed, "has been one of the worst features of non-planning in the second half of the '50s".[3] The building opened in 1964, having taken over four years to complete. Auerbach must have made sketches of the site shortly after the demolition of Westminster Chambers and as the ground works for the new building were beginning, during the latter half of 1959. A newspaper article drew readers' attention towards the short-lived vista that had been opened up, revealing "Old Westminster from a New Angle" (fig. 44). Auerbach did not take the opportunity to include Westminster Abbey in his composition – just as he had marginalised St Paul's Cathedral in an earlier painting (cat. 6). Instead, he focused upon a large pile of sand or rubble, probably reshaping it somewhat and exaggerating its size. Its pyramidal form is also suggestive of the tripod crane-rigs evident in contemporary photographs of the site. Set against the geometric regularity of the office windows in the background, this massed pile spilling out into the formless earthy foreground, with its clutter of building materials, conveys a dramatic contrast between architectural order and its origins in the chaos of demolition.

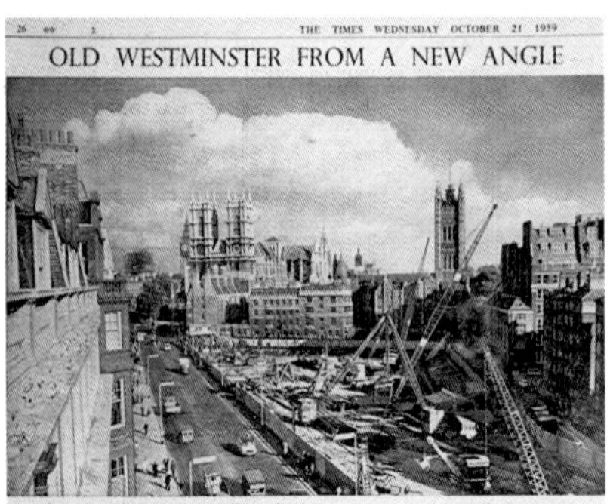

Fig. 44
The construction works for the future Department of Trade building in Victoria Street shortly after the demolition of Westminster Chambers
The Times, 21 October 1959

NOTES

1 Auerbach/Wright interview 2009.
2 Nos. 1–19 Victoria Street.
3 *Architects' Journal*, January 1960, p. 143.

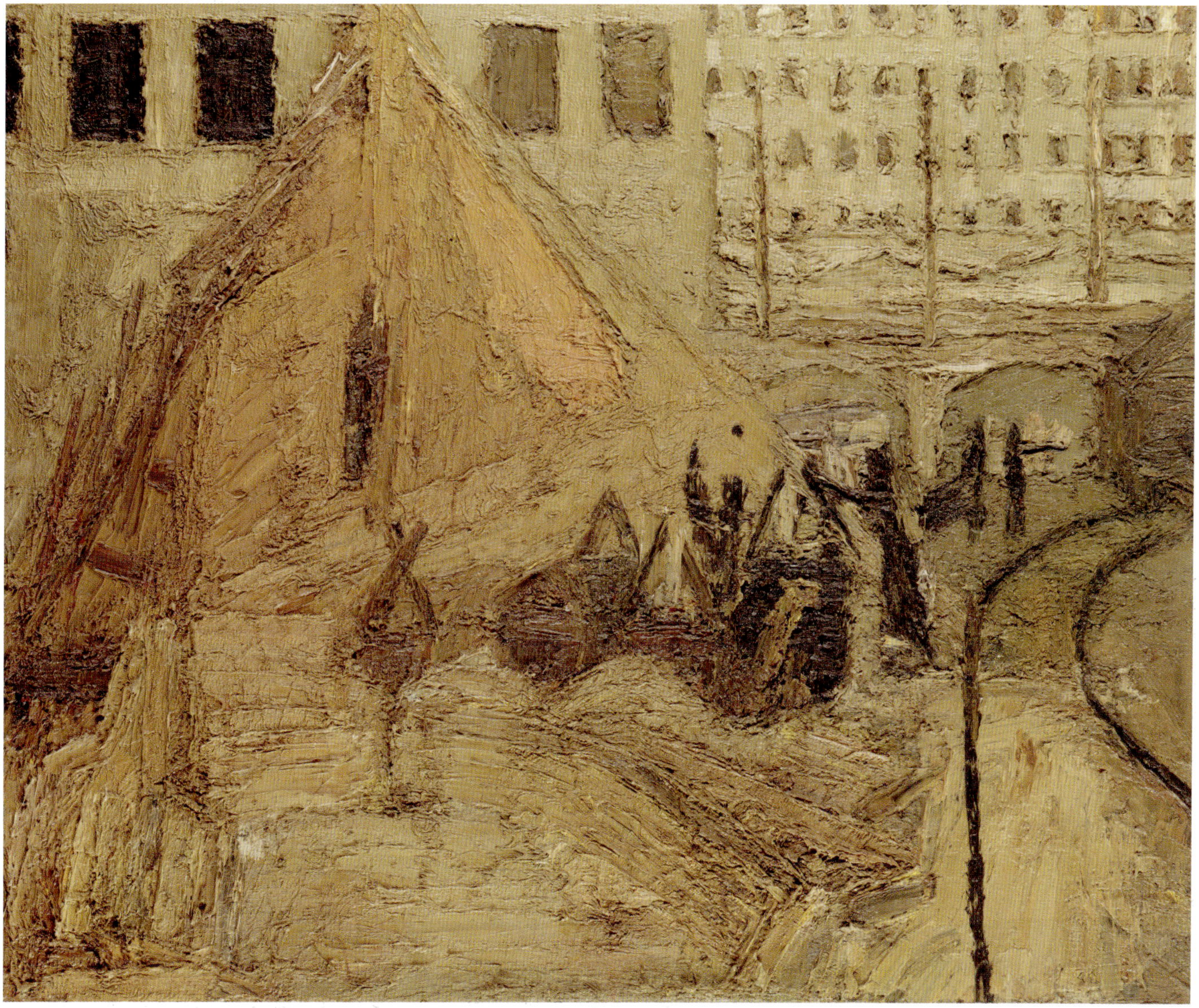

11a

Oxford Street Building Site I

1960

Oil on board
198 × 153.6 cm
Tate, London

11b

Oxford Street Building Site II

1960

Oil on canvas
213.4 × 168 cm
National Gallery of Victoria, Melbourne
Australia

SELECTED EXHIBITIONS

(*Oxford Street Building Site I*):
Frank Auerbach, London, Beaux Arts Gallery, April–May, 1961; *The Hard-Won Image, Traditional Method and Subject in Recent British Art*, Tate Gallery, London, 4 July–9 September 1984; *The Transformation of Appearance, Andrews, Auerbach, Bacon, Freud, Kossoff*, Sainsbury Centre for Visual Arts, University of East Anglia, Norwich, 24 September–8 December 1991
(*Oxford Street Building Site II*):
Frank Auerbach, Beaux Arts Gallery, London, April–May, 1961; *Frank Auerbach*, Hayward Gallery, London, 4 May–2 July 1978, and Fruit Market Gallery, Edinburgh, 15 July–12 August 1978
(*Oxford Street Building Site, Sketches I–III*):
Frank Auerbach, London, Beaux Arts Gallery, April–May 1961

This composition was the only one of the building-site series that Auerbach decided to produce as two versions.[1] Larger, on canvas rather than his typical board, and with a greater clarity of detail than *Oxford Street Building Site I*, *Oxford Street Building Site II* was a self-conscious endeavour to make a major artistic statement. "I remember reading that Picasso's *Demoiselles d'Avignon* had been painted on a lined canvas, because it was his attempt to paint a masterpiece, in that very healthy French tradition where everybody tried to paint a masterpiece in their youth," he recalled. "So I went to Robertson's, which was an art shop in Parkway up the road, and asked them to line a canvas for me. They didn't have the faintest idea what a lined canvas was! They simply stretched two canvases one over the other on the same frame . . . . But it was my attempt to paint a proper big important picture."[2]

Its subject was the important and well-known construction works on the site of London's famous John Lewis department store, which stretched from Oxford Street to Cavendish Square.[3] John Lewis was the only major West End store to be largely destroyed during the Blitz, its flagship nineteenth-century building being directly hit by several bombs in the early hours of Wednesday 18 September 1940. The fires that raged were exacerbated by strong winds, which blew the flames across Holles Street to John Lewis's adjacent building, while a further bomb fell shortly after the fire brigade had arrived, killing three firemen (fig. 45). The store's stoical determination to press on with 'business as usual' saw shop assistants initially setting out the store's wares (including, memorably, the lingerie displays) on the pavement outside whilst usable space was being salvaged from the ruins behind. A highly visible symbol during the war of the 'Blitz spirit', in the heart of London's principal shopping district, the John Lewis store became an equally potent symbol of post-war reconstruction. In 1955 work began on a large new building by Salter

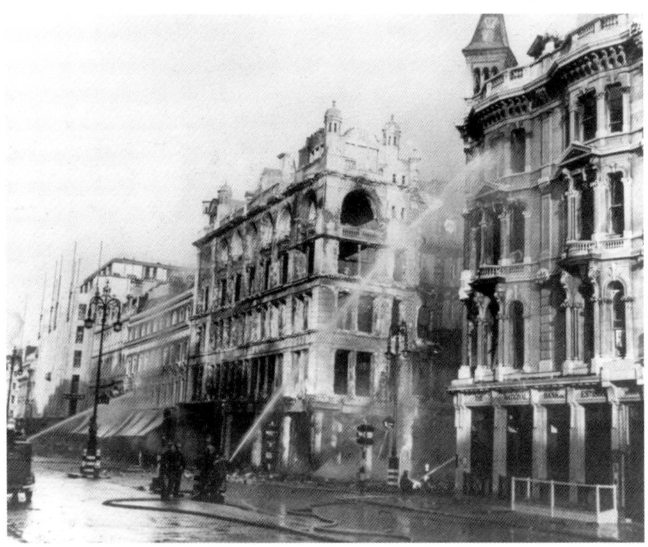

Fig. 45
The fire being extinguished at the John Lewis store on Oxford Street following its bombing during the early hours of 18 September 1940
Photograph: John Lewis Archives

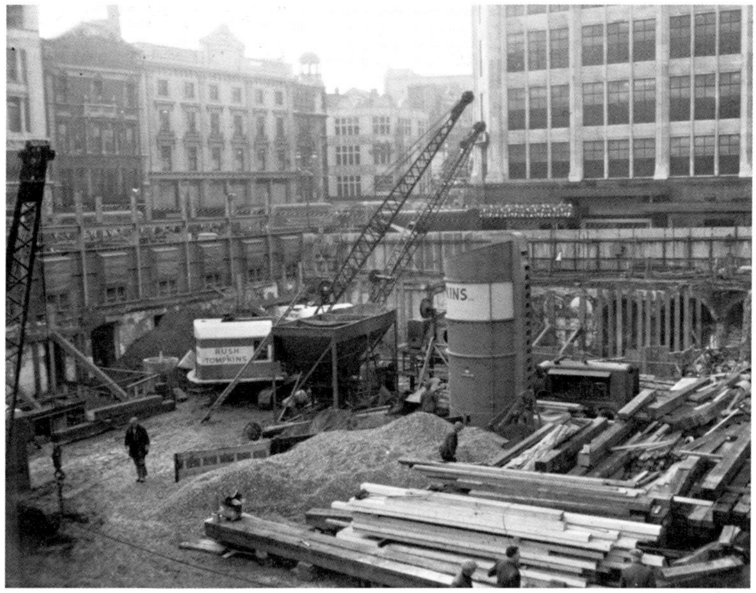

Fig. 46
An area of the new John Lewis store on Oxford Street under construction, 1956
Photograph: John Lewis Archives

& Unwin which would completely replace the devastated nineteenth-century structure and offer a spacious, bright and airy modernist home for the store (figs. 46 and 47). When it opened in 1960 most agreed its design was successful without being particularly imaginative, the *Financial Times* proclaiming it "a well-modelled, honest-to-god twentieth-century store, big in scale, dignified and welcoming."[4] In 1963 Barbara Hepworth's wall-mounted sculpture, *Winged Figure*, was unveiled on the side of the building – the final statement of John Lewis's phoenix-like rising from the ashes of war.

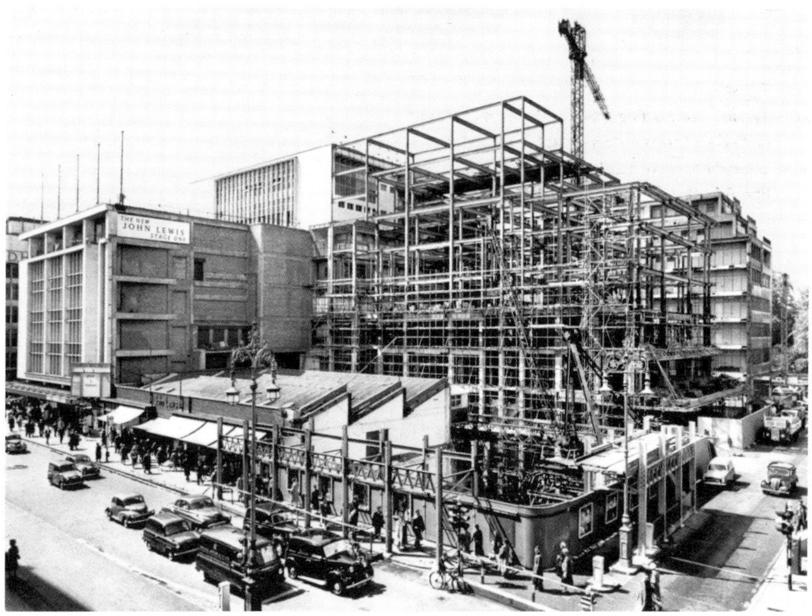

Fig. 47
The new John Lewis store on Oxford Street under construction, *c.* 1957–58
Photograph: John Lewis Archives

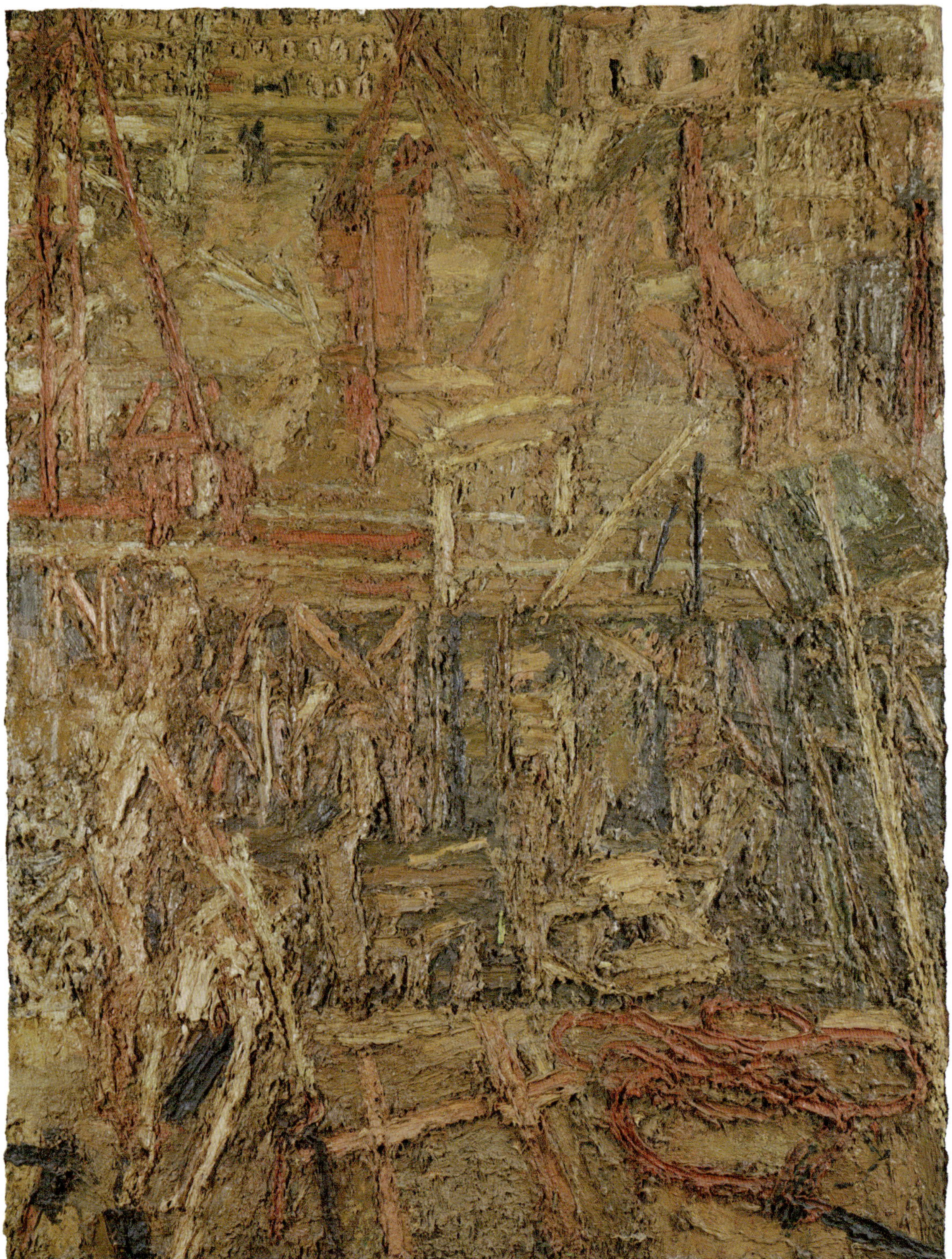

11a

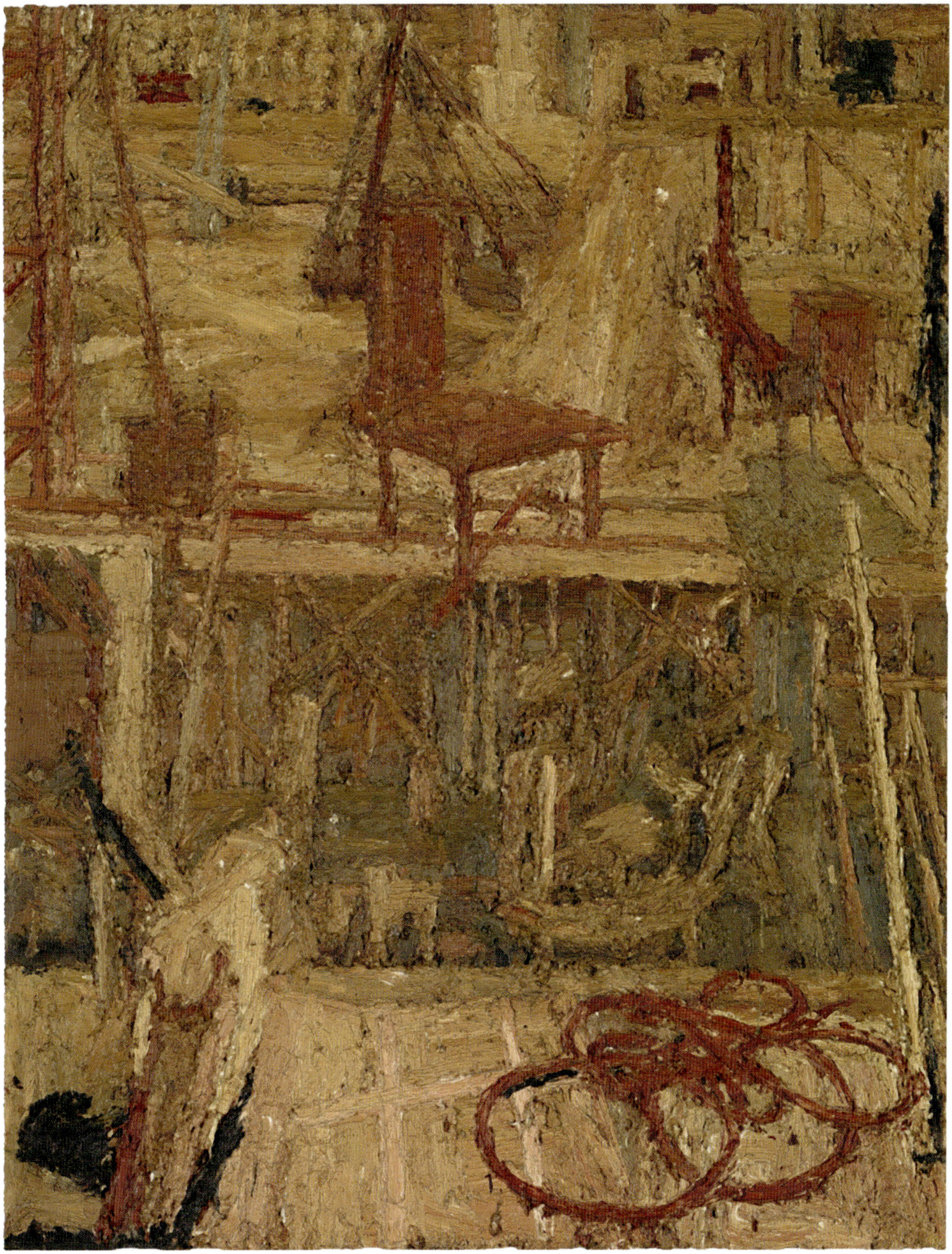

11b

11 (i)

Oxford Street Building Site, Sketch I

c. 1959–60

Oil on board
50.8 × 35.5 cm
Private collection

11 (ii)

Oxford Street Building Site, Sketch II

c. 1959–60

Oil on board
48.6 × 40.6 cm
Private collection

11 (iii)

Oxford Street Building Site, Sketch III

c. 1959–60

Oil on board
53 × 39.5 cm
Pallant House Gallery, Chichester

When Auerbach made his sketching trips, quite possibly a year or more before he embarked upon the paintings, demolition and rebuilding work were progressing in phases to maximise the shop's ability to keep trading in areas of temporary accommodation on the site. His chosen view was a scene of considerable activity looking across a shored-up excavation being worked on by several men, with machinery and equipment strewn across the site. Of all the paintings in the series, the largest number of related oil sketches (cat. 11 (i–iii)) and drawings survive for this subject. Although an incomplete record, they give a clear sense of the central role that sketching played in Auerbach's practice and remind us that each of his building-site paintings went through similar stages of preparation. The four Tate drawings (cat. 11 (iv–vii)) demonstrate how Auerbach sketched a particular view and then focused upon specific aspects of it in subsequent rapid sketches. The squared-up drawing (cat. 11 (viii)) gives written colour notes and records other details such as a 'no entry' sign on the right-hand side. Despite being squared up the drawing's composition is not faithfully carried through to the finished paintings, which actually relate somewhat more closely to the Tate sketches. It was typical for Auerbach to amalgamate different elements of his sketched views during the painting process, only taking from drawings what was relevant for the development of his composition: "If they became useless or didn't convey any information, I might take them down and pin up something else," he explained.[5] The three figure drawings (cat. 11 (ix–xi)) were an unusual extension of this practice. He recalls wanting to put figures into the foreground of the painting and at the end of a life class he was teaching at Camberwell School of Art he asked the model to adopt workmen's poses, which he quickly sketched in a notebook. These figures, one bending and one standing, would become a major element of the lower left-hand corner of the finished paintings.

Auerbach began work on *Oxford Street Building Site II* in 1959 and started *Oxford Street Building Site I* a few months later, working on both compositions simultaneously for a period. He completed the Tate work first, around November of 1960, and did not finish the large canvas until the end of the year.[6] The following year both paintings, together with all three oil-sketches, were exhibited at his Beaux Arts Gallery solo show. Seen as a group the five paintings chart Auerbach's exploration and gradual resolution of the composition as he developed its distinctive banded structure, formed round the darkly painted void of the excavation running through the centre. His relatively high viewpoint and lack of a horizon line have the effect of flattening the picture plane, asserting the immediacy of the paintings' heavily worked surfaces to the point of disorienting the viewer. The upper section departs from that seen in the surviving pencil sketches and includes tripod rigs among an assortment of scaffolding poles, beams, platforms and other equipment (no doubt taken from drawings that are no longer extant). Auerbach renders this riot of building equipment as deeply gouged lines in the paint surface,

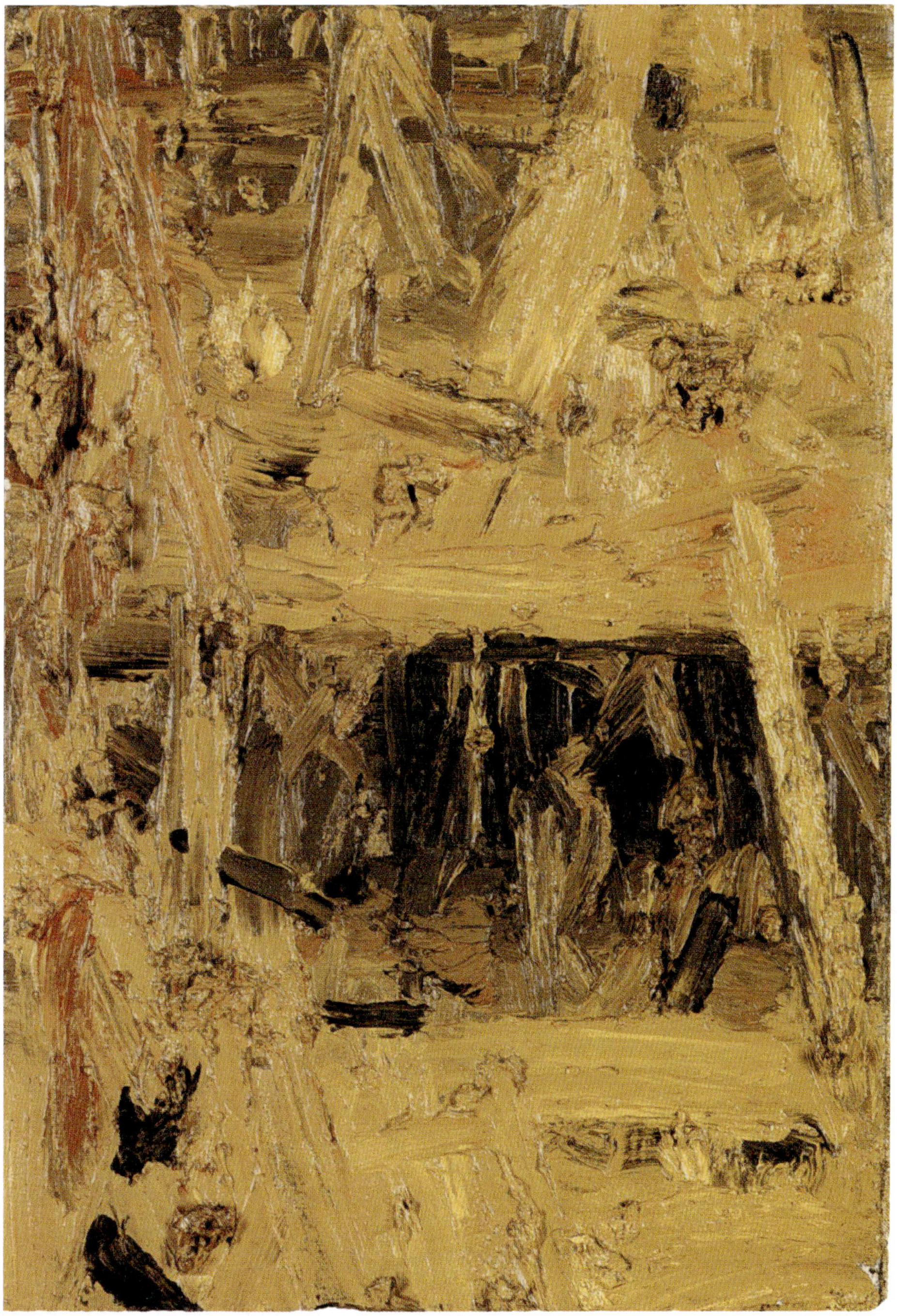

11 (i)

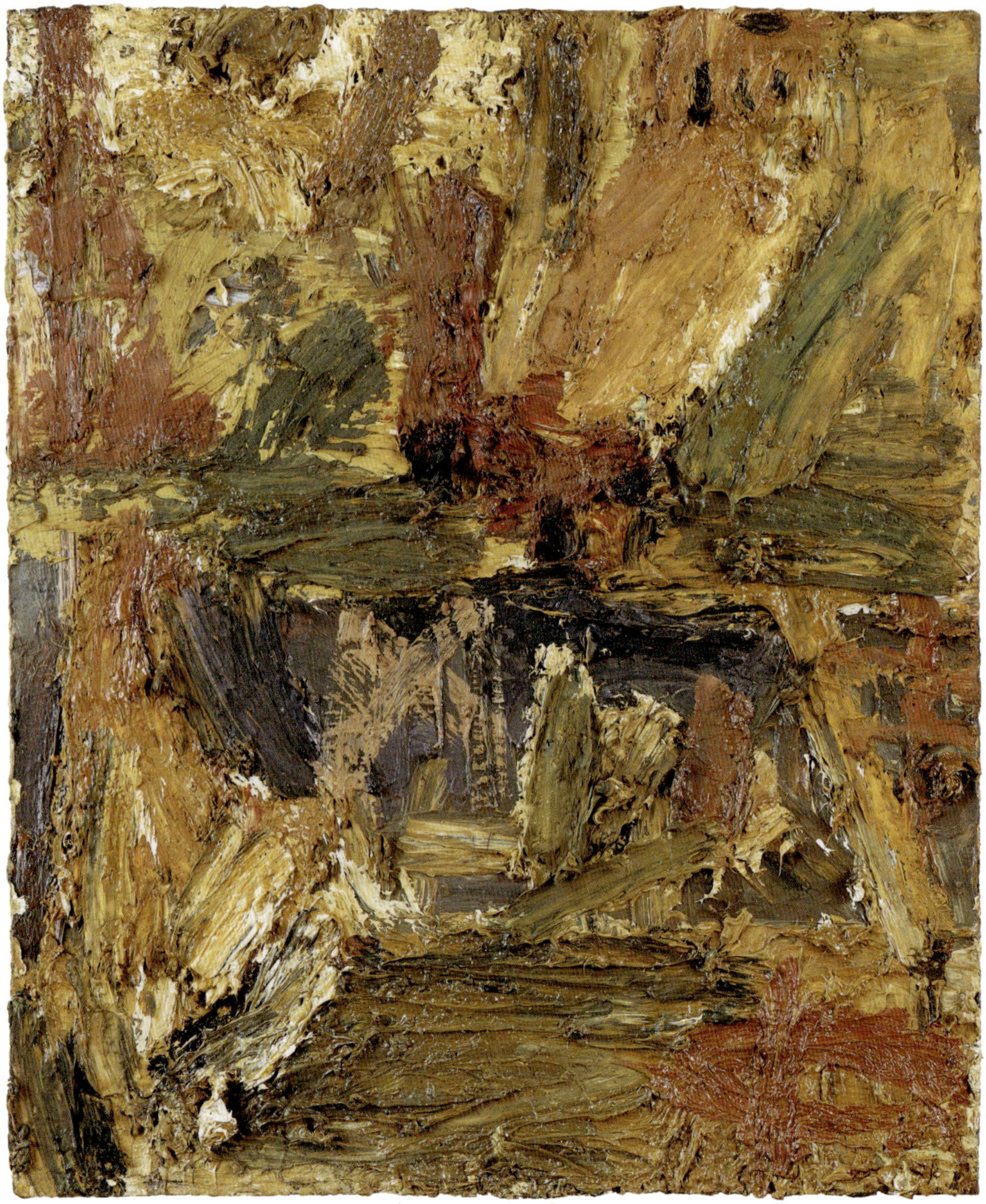

11 (ii)

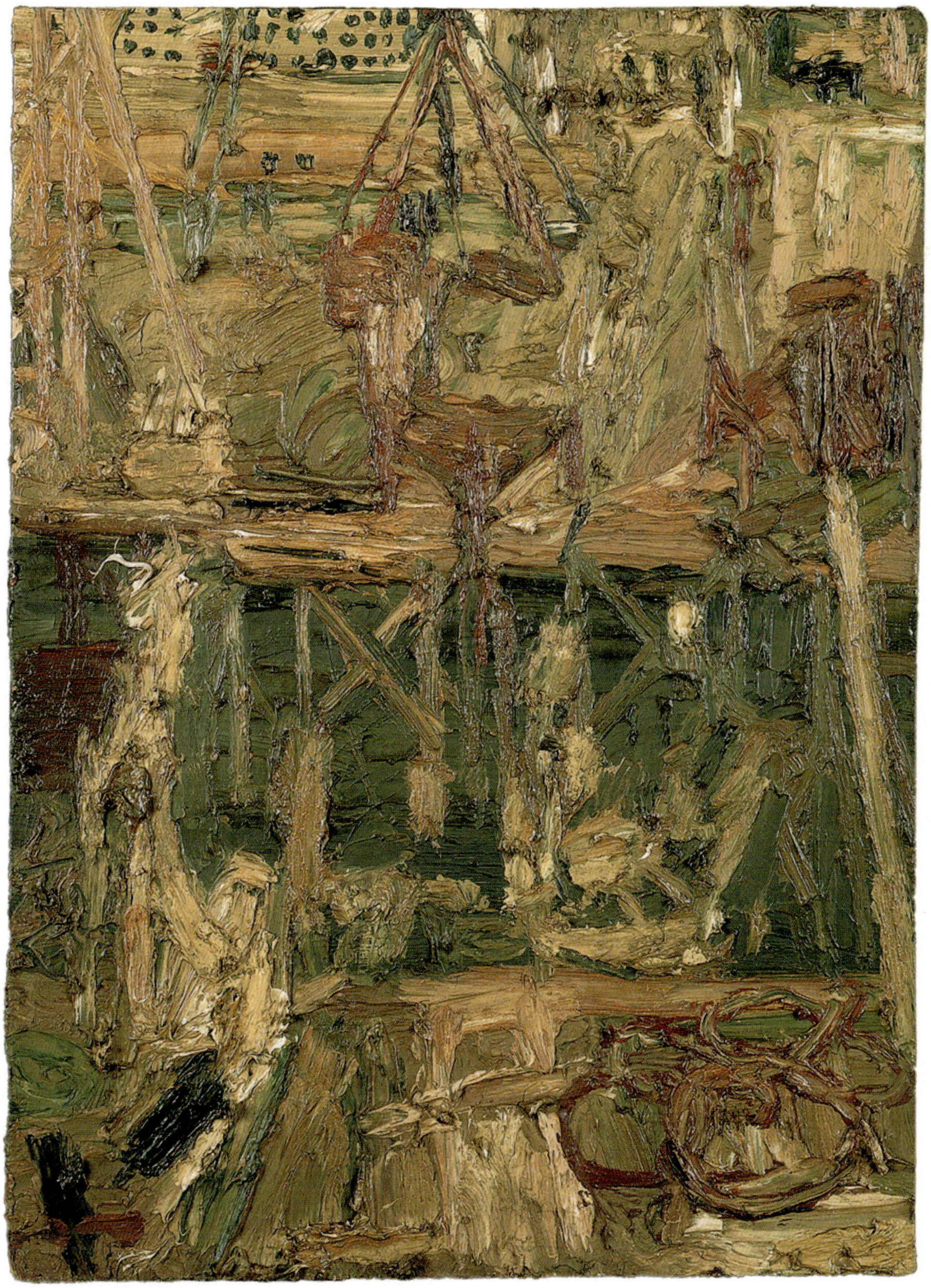

11 (iii)

11 (iv–vii)

Studies for Oxford Street Building Site

c. 1957–59

Pencil on paper
16.8 × 21.9 cm (v, vi, viii)
21.9 × 17.2 cm (vii)
Tate, London

scarring and disrupting the three broad bands of colour which act as a ground layer for his urgent and fierce markmaking. His need to work on at least three oil sketches as he executed the two principal paintings suggest a particularly restless period rehearsing the composition – the small scale of his sketching boards allowing him greater potential to paint freely and spontaneously. The Tate painting retains much of this sense of immediacy and spontaneity whereas in the Melbourne picture Auerbach's desire for greater formal structure and clarity is asserted. In both works the insistent rigid linearity is relieved in the lower right-hand corner by the gracefully looping form of a coil of cable or rope, taken from a small detail in one of the pencil sketches (cat. 11 (vii)). This unexpected and almost playful aside finally secures the balance of the composition. The ever-shifting elements of the building site itself, with workmen, equipment, earth and structures being constantly reconfigured as the construction progresses, is a potent analogue to Auerbach's urgent mark-making and shuffling of elements from his drawings in pursuit of what he has described as a "formal grandeur".[7]

The two Oxford Street paintings are a vivid and monumentally scaled expression of Auerbach's attempts to renew and recast the language of painting. Displayed at the 1961 exhibition, the paintings and their oil-sketches helped to further critical understanding of his approach, notably in an important and influential article by David Sylvester,[8] Auerbach's foremost champion at this time. Both paintings were bought for national collections (Tate Gallery and the National Gallery of Victoria, Melbourne, respectively), which represented a significant stage in the establishment of Auerbach's reputation.

NOTES

1 Auerbach produced a replica of cat. 3 for the Royal College of Art (see fig. 35) but it was produced two years later as a record of the original and not as a second version in the way he conceived cat. 11b.
2 Auerbach/Wright interview 2009.
3 In a letter (30 May 1960) to the Tate Gallery regarding the location of his subject, Auerbach stated that the drawings were made from "a building site which stretched from Oxford Street to Cavendish Square, but I cannot be certain whether it formed the foundation of the John Lewis extension or the building next to it": Tate Archive. The John Lewis construction works covered this entire block. The site was complex and developed in different phases, making works look somewhat disconnected, but Auerbach's subject is almost certainly an area of the John Lewis site. John Lewis sold off their adjacent plot on the opposite side of Holles Street in 1958 and construction of a new building for what would become the London College of Fashion commenced towards the middle of 1959; given its late date, this seems less likely to have formed part of Auerbach's sketches.
4 *Financial Times*, Monday 24 October 1960.
5 Auerbach/Wright interview 2009..
6 Auerbach gives the details of this sequence in his 1961 letter to the Tate Gallery, cited in note 2.
7 Auerbach/Wright interview 2009.
8 Sylvester 1961.

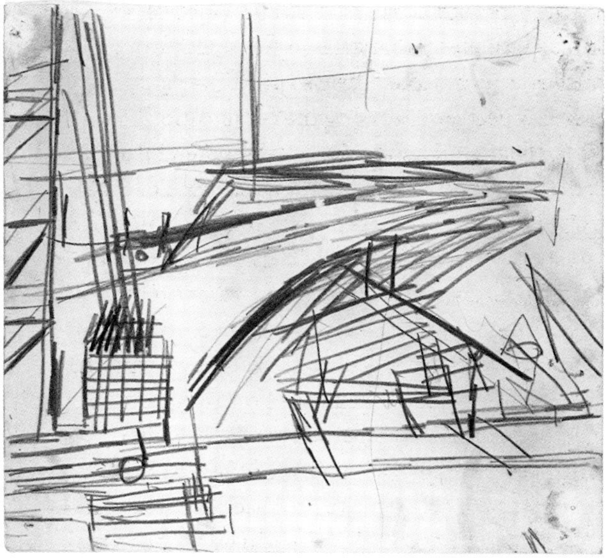

11 (iv)

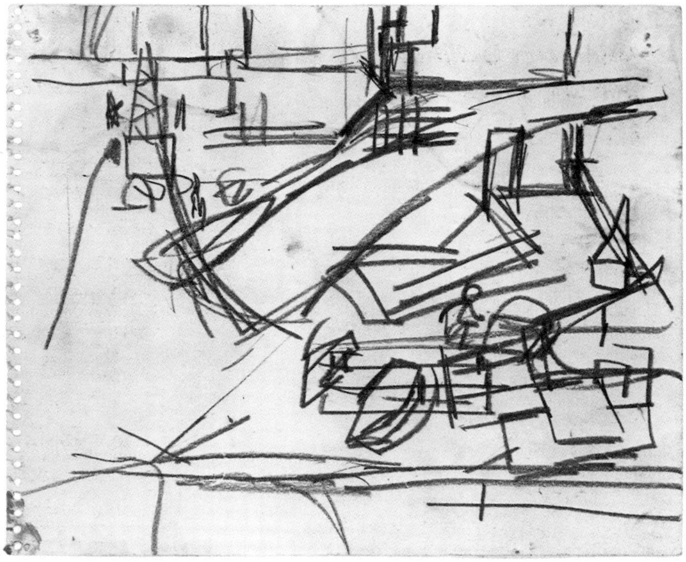

11 (v)

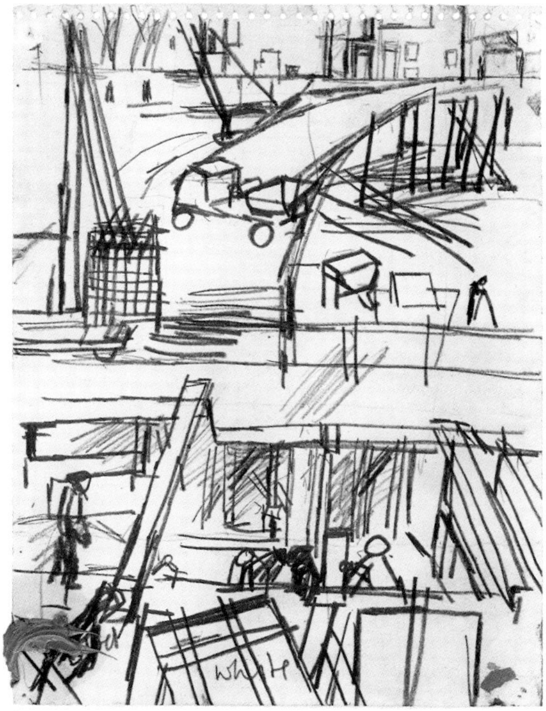

11 (vi)

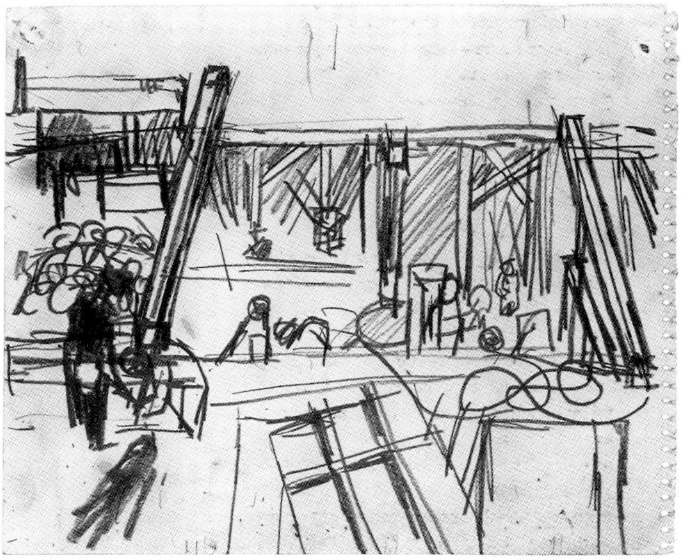

11 (vii)

11 (viii)

Study for Oxford Street Building Site

c. 1957–59

Pencil on paper (squared up in red pastel)
25.9 × 17.9 cm
Private collection

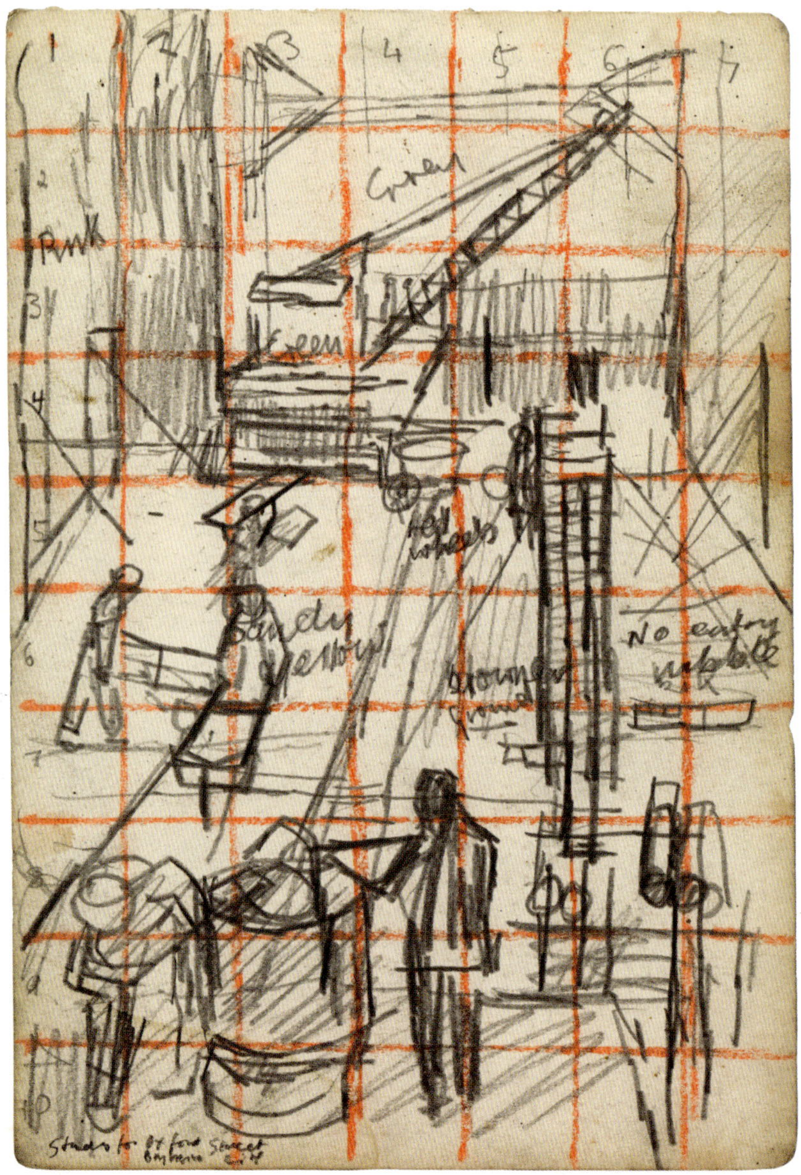

11 (ix–xi)

Figure studies for Oxford Street Building Site

c. 1958–59

Pencil on paper
20.3 × 16.3 cm (each sheet)
Private collection

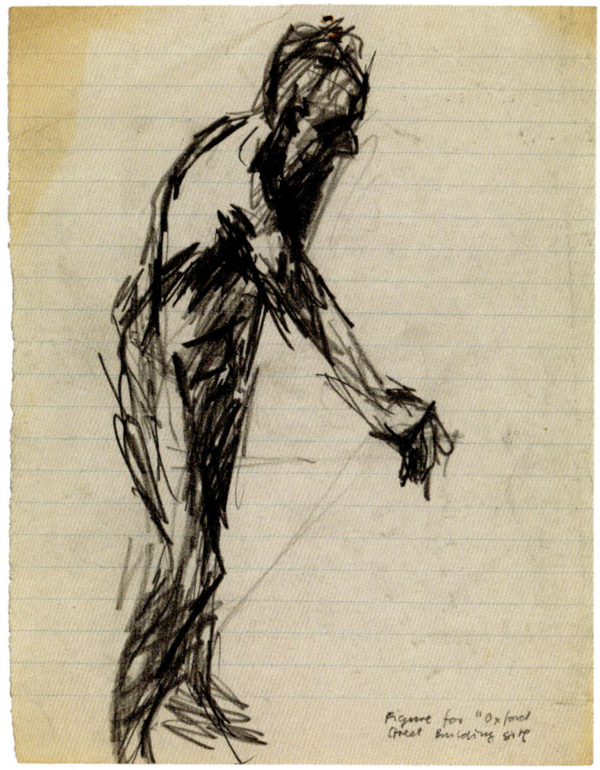

11 (ix)

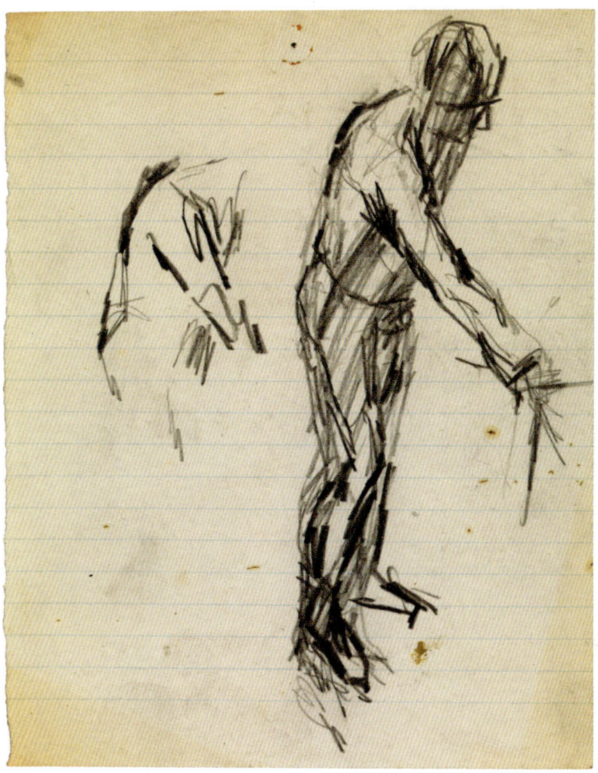

11 (x)

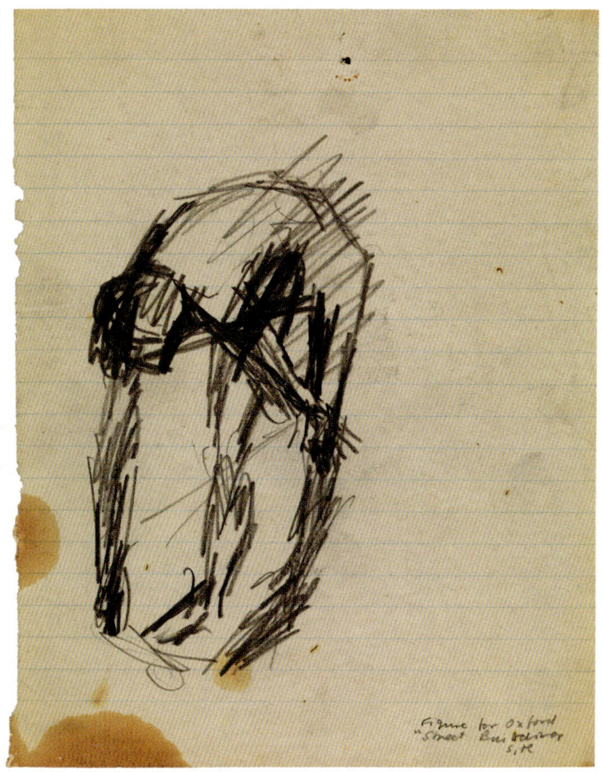

11 (xi)

12

Maples Demolition, Euston Road

1960

Oil on board
148.6 × 153.7 cm
Leeds City Art Gallery

12 (i)

Study for Maples Demolition, Euston Road

c. 1960

Pencil on paper
25.4 × 25.4 cm
Private collection

SELECTED EXHIBITIONS

(*Maples Demolition, Euston Road*):
Frank Auerbach, Beaux Arts Gallery, London, April–May 1961; *The 1961 Pittsburgh International Exhibition of Contemporary Painting and Sculpture*, Carnegie Institute, Pittsburgh, 27 October 1961–7 January 1962; *British Painting in the Sixties, Section One*, Tate Gallery, London, 1–30 June 1963; *Profile III, Englische Kunst der Gegenwart*, Städtische Galerie, Bochum, 19 April–7 June 1964; *Frank Auerbach*, Hayward Gallery, London, 4 May–2 July 1978, and Fruit Market Gallery, Edinburgh, 15 July–12 August 1978; *Jewish Art*, Art Gallery and Museum, Glasgow, 5 September–7 October 1979; *British Art in the 20th Century, The Modern Movement*, Royal Academy of Arts, London, 15 January–5 April 1987, and Staatsgalerie, Stuttgart, 9 May–9 August 1987; *100 Years of Art in Britain*, Leeds City Art Gallery, 4 October 1988–15 January 1989; *British Contemporary Art 1910–1990, Eighty Years of Collecting by the Contemporary Art Society*, Hayward Gallery, London, 3 December 1991–19 January 1992

By the end of the nineteenth century Maple & Co. could bill themselves as "The Largest Furnishing Establishment in the World" and could count the Winter Palace in St Petersburg and numerous British Embassies among their many prestigious clients. Their enormous flagship store occupied an entire block, imposingly positioned on the corner of Euston Road and Tottenham Court Road with extensive window displays stretching along both thoroughfares as far as Gower Street and Grafton Street. Maple's could justifiably claim to be a major London landmark and was a popular tourist destination in its own right. Its vast stock of thousands of furnishing items in every imaginable style, from dining-room suites to stair rods, was an encyclopaedia of Victorian and Edwardian taste. This monument to British Imperial commerce was razed to the ground during the night of 16 April 1941 by German bombing, and almost the entire Maple's complex was reduced to a chaos of rubble and twisted steel girders (fig. 48). Rebuilding of the site was underway by 1950 and would continue for more than ten years, after Maple's had drastically reduced the size of their new premises and sold off the larger part of the plot to Laing & Co. for the development of a large office complex (fig. 49). The deal was indicative of the office-building boom of the 1950s and 1960s and the rise of speculative developers, who rapidly changed the commercial and physical landscape of London.

Auerbach's *Maples Demolition* speaks of the experience of seeing areas of London laid bare by the Blitz and then further dismembered prior to redevelopment. It is the only work in the series specifically to be

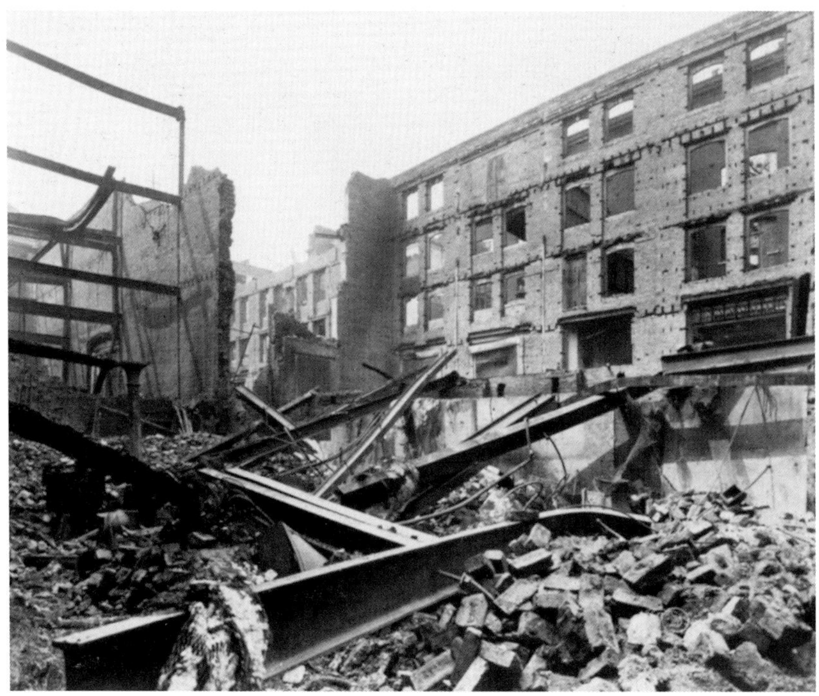

Fig. 48
The Maple & Co. building shortly after being almost completely destroyed by bombing on 16 April 1941
Photograph: Maple & Co. papers, V&A archive

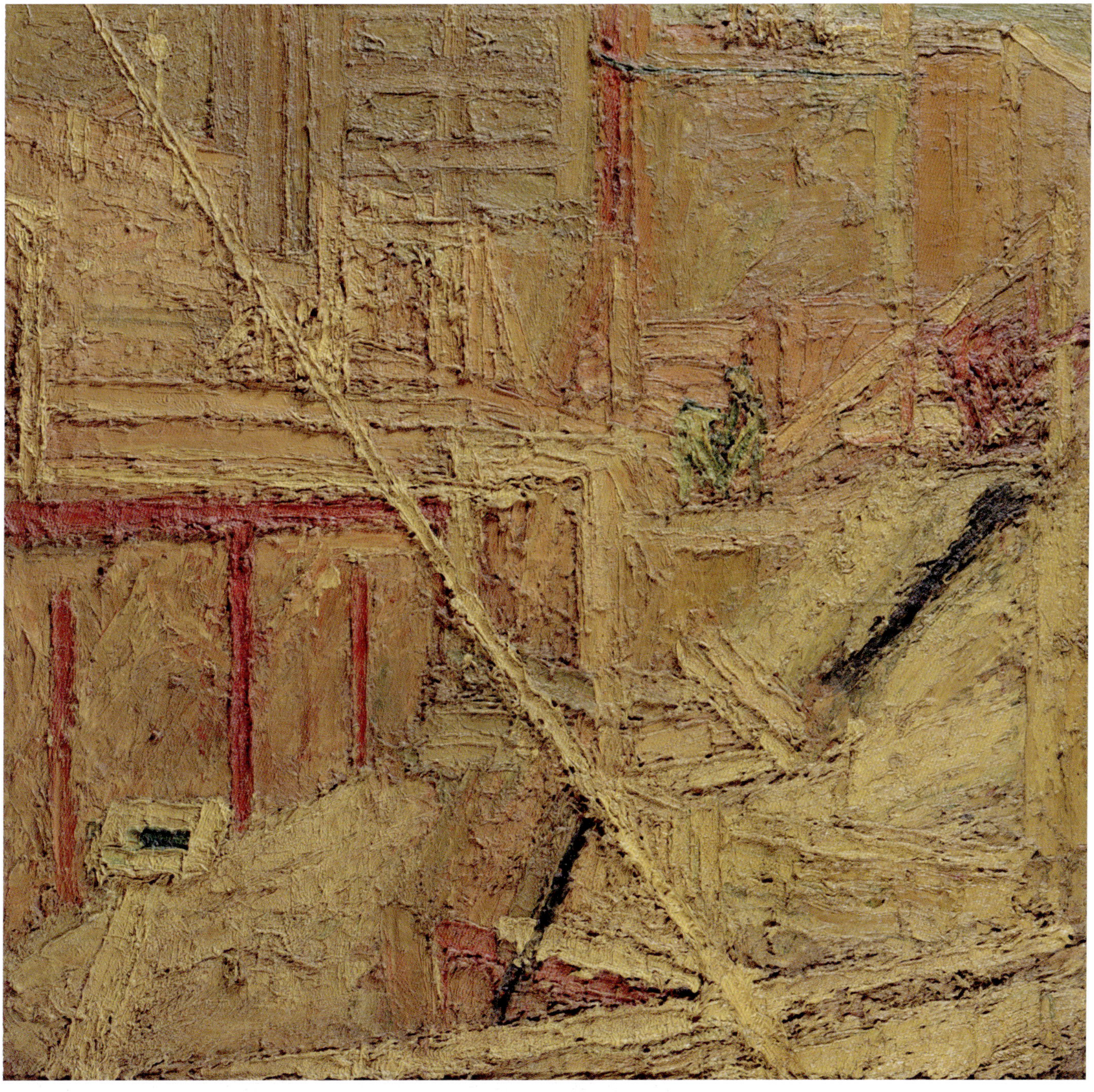

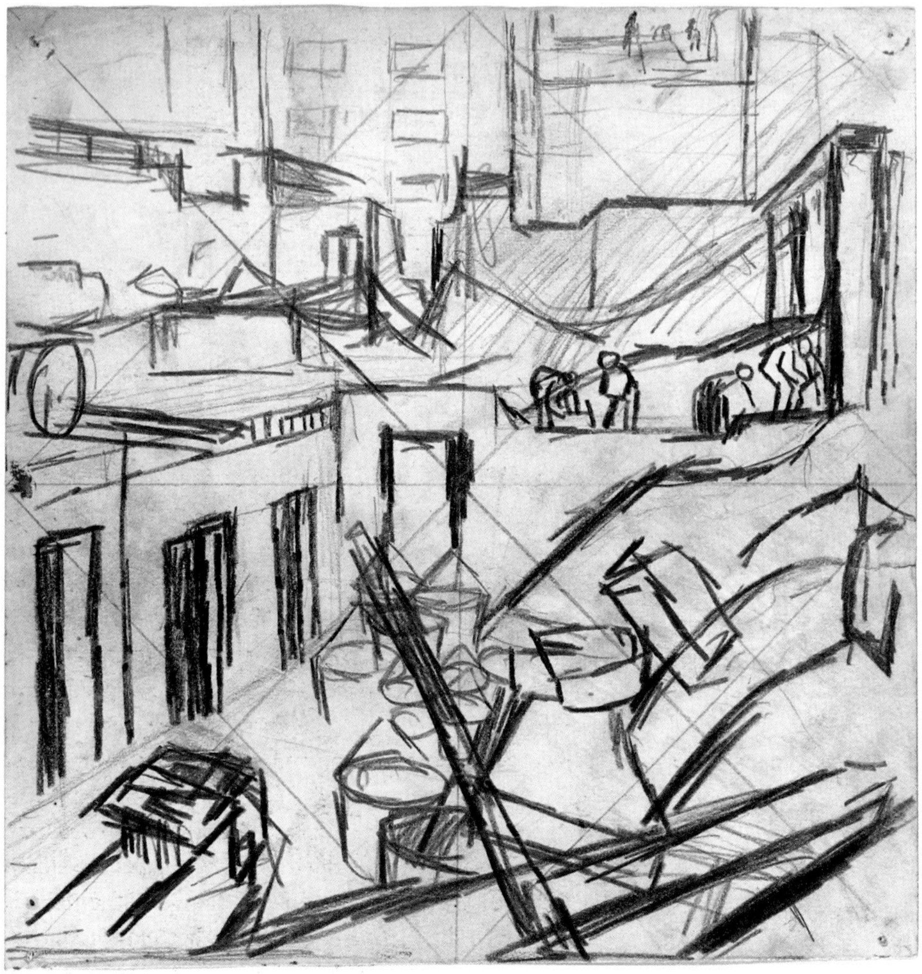

12 (i)

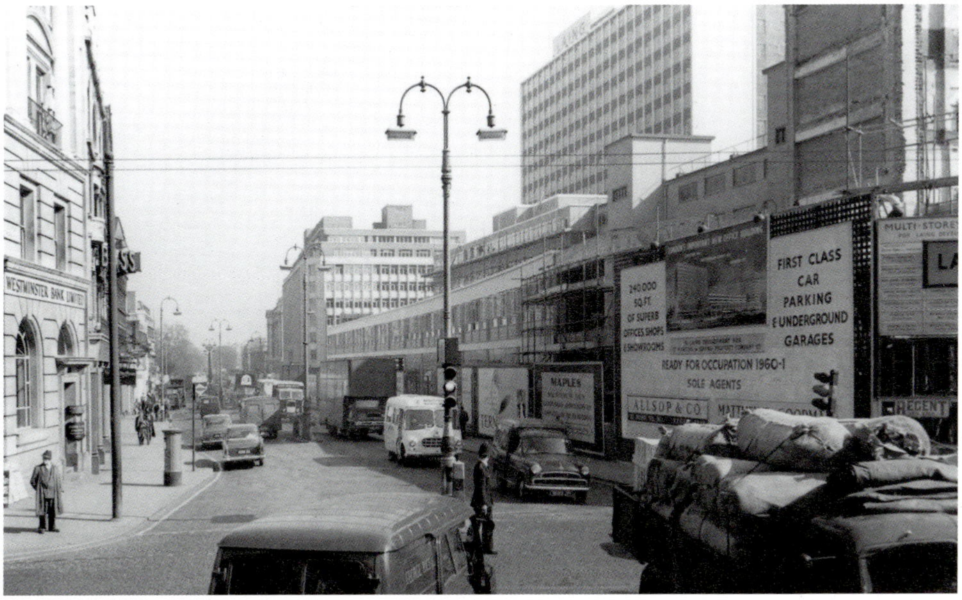

Fig. 49
The construction works on the former Maple & Co. site at the junction of Euston Road and Tottenham Court Road, at an advanced stage, 1961. Advertisements for the Laing-developed office complex and the new, smaller Maple's store can be seen to the right
Photograph: City of London, London Metropolitan Archives

identified as a demolition site, the title cuing us to this sense of a double destruction inherent in the post-war rebuilding of the city. The surviving drawing is a rare example of a squared-up full compositional study for a painting. It depicts a tangle of beams and girders with rubble and equipment strewn around and framed in the background by the broken outlines of ruined walls. Groups of workmen are discernible, with two men drilling and pickaxing towards the centre of the image. Many of these details are preserved in the painting but with various parts amplified to different degrees. Auerbach recalls trying "to extend each element of the composition until it balked or [was] stopped by another so that the energy [was] as high as possible".[1] The ultimate expression of this was his decision to extend the diagonal line of the beam in the foreground right across the image. This slashing line is at once constructive and destructive, intensifying and securing the composition whilst at the same time defacing it like scoring to cancel an etching plate.

Maples Demolition was exhibited alongside the Oxford Street building-site paintings (cat. 11a and 11b) at Auerbach's 1961 Beaux Arts Gallery show and was singled out by several commentators as being one of his most powerful and successful paintings. His dealer, Helen Lessore, thought it to be the masterpiece of the building-site series,[2] and the critic for *The Times* celebrated it as being "majestic in its simplicity" and "as massive an identification of space, substance and structure as Mr Auerbach has yet painted".[3] Later that year the painting was included (alongside works by a number of major painters working in America during the period, including Mark Rothko and Willem de Kooning)[4] in the Pittsburgh International Exhibition at the Carnegie Institute, helping to further Auerbach's international exposure.

NOTES

1 Auerbach/Wright interview 2009.
2 Letter from Helen Lessore to Sir William Coldstream, 7 October 1960 (Beaux Arts Gallery archive).
3 Thompson 1961.
4 Auerbach had also exhibited in the 1958 Pittsburgh International Exhibition, showing *E.O.W. Nude*, 1953–53 (fig. 27).

13

Rebuilding the Empire Cinema Leicester Square

1962

Oil on board
152.5 × 152.5 cm
Private collection

SELECTED EXHIBITIONS

Frank Auerbach, Beaux Arts Gallery, London, September 5–October 12, 1963; *Frank Auerbach*, Marlborough-Gerson Gallery, New York, September– October 1969; *Frank Auerbach, Paintings and Drawings 1954–2001*, Royal Academy of Arts, London, 14 September–12 December 2001

"I remember passing the Empire Cinema in Leicester Square and looking in ... and doing a drawing, and I'm not even certain whether I wasn't warned that I had no right to be there, but I did the drawing with a very considerable feeling of urgency. Time had passed on and I had a feeling that this was a very efficient building site and they were doing their damndest to get it done as quickly as possible. I knew what I saw there would no longer be there perhaps a fortnight later .... The composition in a way was a gift ... you looked through and saw something marvellous .... At the time I would have been able to tell you all sorts of things but ... it's a bit like going to the dentist, it's all so stressful that once it's over you tend to forget the whole business."[1]

Auerbach had chanced upon the interior reconstruction of London's famous Empire Cinema in Leicester Square. The cinema had closed on 28 May 1961, after a 76-week run of *Ben Hur*, amidst growing criticism of the theatre's infrastructure, which was struggling to do justice to the ever-more spectacular and technically advanced Hollywood productions. The frenetic activity Auerbach witnessed reflects the fact that the builders were working against the clock, with barely eighteen months to finish the complete rebuilding of the interior to designs by George Coles. Coles stripped out the ornate nineteenth-century interior to make way for a sleek modern design, extending the former circle area into one large new floor with a rear stepped section, providing 1,330 seats. The latest Perlux screen was fitted; measuring approximately 30 × 60 feet, it was the largest in London, with advanced projection and stereo equipment completing the system. Below, the old stalls area was transformed into a large ballroom, enabling the Empire's new owners, Mecca, to offer a dance venue alongside their state-of-the-art modern cinema. The new Empire was unveiled on 19 December 1962, premiering the lavish MGM circus musical, *Jumbo*, starring Doris Day.

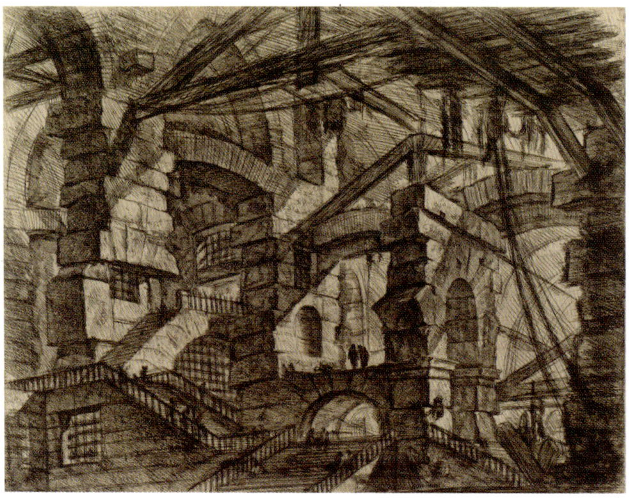

Fig. 50
Giovanni Battista Piranesi, *The Gothic Arch*, from *Carceri*, plate XIV (first edition)
c. 1749–60
Etching and engraving
41.5 × 51.3 cm (plate mark)
The Courtauld Gallery, London

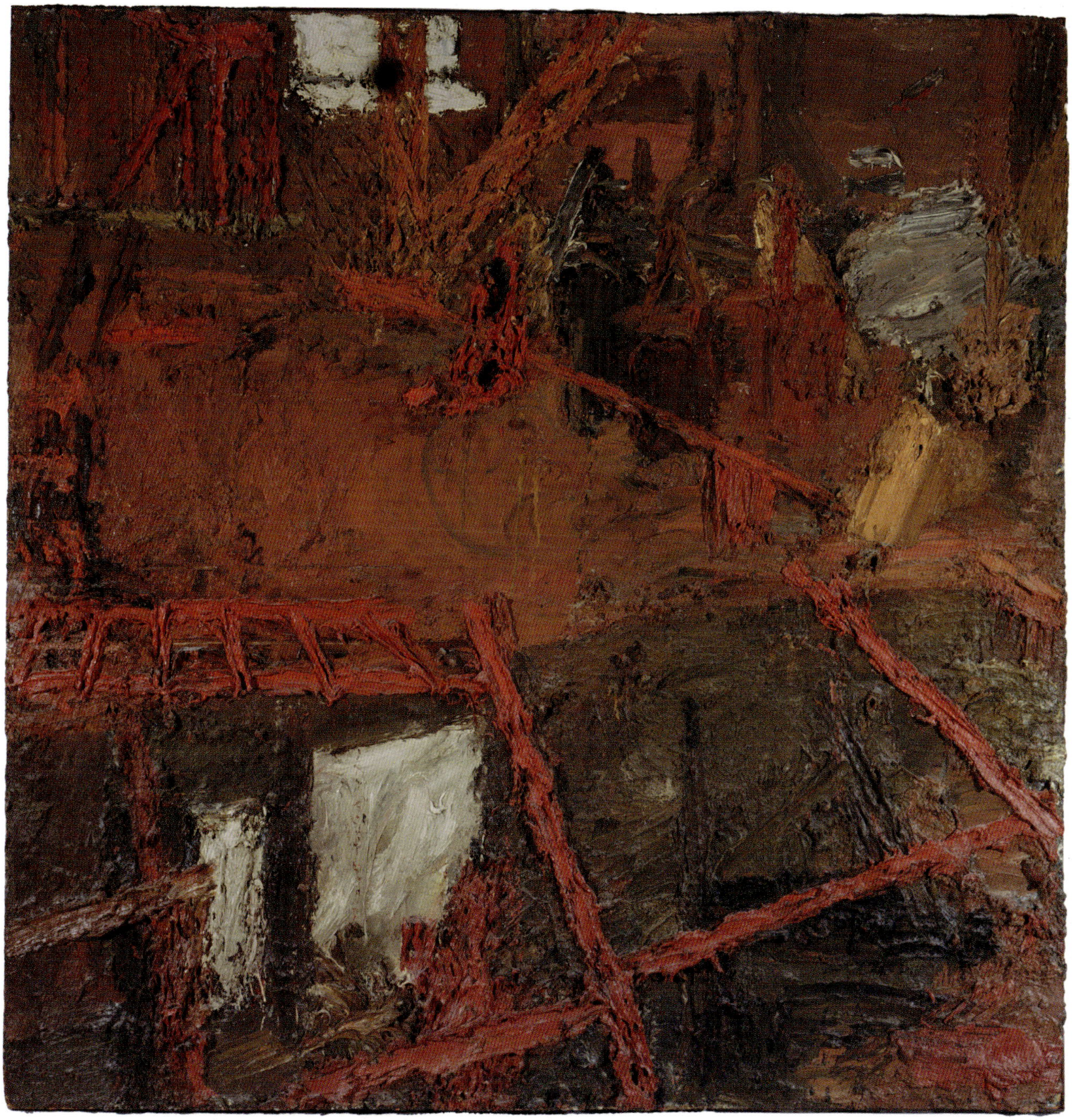

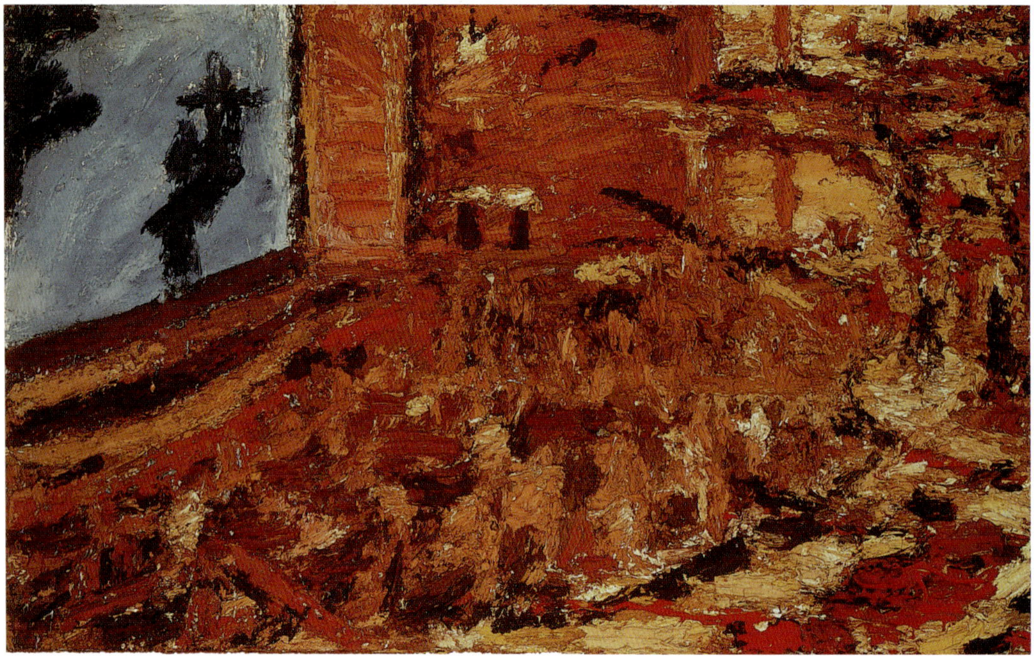

Fig. 51
Frank Auerbach, *The Gaumont Cinema, Camden Town* 1963
Oil on board, 89 × 146 cm
Private collection

Auerbach translates his fraught experience of peering in at this dramatic scene as the building works forged ahead into a painting of exceptional energy and visual excitement. The skeletal framework of red painted beams defines the gaping chasm in the foreground – as if the viewer were teetering over the edge. The shafts of light exacerbate this vertiginous sensation and enhance the experience of the work as an enclosed space which compresses the energy of the composition. The painting prompts comparison with Piranesi's *Carceri* series, to which David Bomberg had introduced Auerbach more than ten years earlier (see fig. 50). Piranesi's disorienting images of labyrinthine imaginary prisons made a deep impression upon the young Auerbach, which resonates throughout his building-site paintings, but it is this interior scene, with its part-demolished floor levels and spatial ambiguities, unleashed from the certainties of architectural order, which most powerfully demonstrates Auerbach's deep understanding of the *Carceri*'s aesthetic. His use of red pigment marks Auerbach's increasing ability to move on from earth tones to more expensive coloured paints, and here intensifies the fiery, almost hellish, atmosphere of the scene. This sense of compressed and heated intensity also characterises the only other cinema painting of his career, depicting a screening at his local Gaumont Cinema in Camden Town (fig. 51) – both works being exhibited at his last Beaux Arts Gallery show in 1963.

Rebuilding the Empire Cinema would prove to be the last building site Auerbach painted. From this point onwards he has focused almost exclusively upon landscapes chosen from his immediate locality of Camden Town and Primrose Hill and upon studio-based figure subjects. Having been stimulated by building sites for more than a decade, Auerbach felt he had exhausted the subject. The end of the series was aligned to the fact that for him the Empire Cinema construction works signalled a change in the type and character of London's building sites, as the city became more functional, ordered and regularised. It marked an end to his post-war experience of London as a scene of "survivors scurrying among a ruined city".[2]

NOTES

1 Auerbach/Wright interview 2009.
2 *Ibid.*

Bibliography

What follows are selected lists of material that has been consulted in the preparation of this catalogue. For more comprehensive bibliographies relating to Auerbach see: Mathis 1993; *Frank Auerbach: Recent Work* (exhibition catalogue), Marlborough Fine Art, London, 1990; Lampert 2001; and Feaver 2009. Exhibition histories for each of the building-site paintings are listed in the individual entries. Annotated Gallery copies of catalogues for Auerbach's Beaux Arts Gallery exhibitions between 1956 and 1963 are held in the Beaux Arts Gallery Archive in the collection of Mr Henry Lessore.

Selected Writings, Interviews and Statements

(In chronological order)

AUERBACH 1958
Frank Auerbach, 'Seven Portraits', *ARK*, 23, 1958

AUERBACH 1959
Frank Auerbach, 'Fragments from a Conversation', in D. Wright and P. Swift (eds.), *X: A Quarterly Review*, vol. 1, no. 1, London 1959

AUERBACH/BATTYE 1971
'Frank Auerbach talks to Christopher Battye', *Art and Artists*, January 1971

AUERBACH/LAMPERT 1978
'A Conversation with Frank Auerbach', interview by Catherine Lampert in *Frank Auerbach* (exhibition catalogue), Hayward Gallery, London, and Fruit Market Gallery, Edinburgh, 1978

AUERBACH 1981
Frank Auerbach, 'Foreword', in *Late Sickert: Paintings from 1927–1942* (exhibition catalogue), Hayward Gallery, London, 1981

AUERBACH/CORK 1985
'In Pursuit of Change', Frank Auerbach in conversation with Richard Cork, broadcast BBC Radio 3, 18 November 1985

AUERBACH/BUMPUS 1986
'Frank Auerbach', interview with Judith Bumpus, *Art and Artists*, June 1986

AUERBACH 1987
'Frank Auerbach' (interview), *Bamboo*, vol. 14, Autumn 1987

AUERBACH/PEPPIATT 1987
'Frank Auerbach Going Against the Grain', interview with Michael Peppiatt, *Art International*, Autumn 1987

AUERBACH/CORK 1988
'Frank Auerbach', interview with Richard Cork, *Art & Design*, vol. 4, no. 9/10, 1988

AUERBACH/PEPPIATT 1998
'Frank Auerbach', interview with Michael Peppiatt, *Tate*, no. 14, Spring 1998

AUERBACH/GREIG 1998
'Paint or Die: Meeting Frank Auerbach', interview with Geordie Greig, *Modern Painters*, vol. 11, no. 3, Autumn 1998

AUERBACH/GREIG 1998
'Hidden Talent', interview with Geordie Greig, *Time Magazine*, 12 December 1998

AUERBACH/GAYFORD 2001
'Frank Auerbach, A Master at Seventy', interview with Martin Gayford, *The Daily Telegraph*, 1 September 2001

AUERBACH/TUSA 2001
'The John Tusa Interview with Frank Auerbach', broadcast BBC Radio 3, 7 October 2001

To the Studio 2001
'Frank Auerbach, To the Studio', Hannah Rothschild (director), Jake Auerbach (producer), Jake Auerbach Film Productions, 2001 (first broadcast Omnibus, BBC 2, 10 November 2001)

AUERBACH/WRIGHT 2009
'Frank Auerbach – The Building-Site Paintings', interview with Barnaby Wright, 15 January 2009; transcript and recording held in The Courtauld Gallery Archives

Selected Exhibition Reviews (1954–63)

(in chronological order)

SYLVESTER 1954
David Sylvester, 'Round the London Galleries', *The Listener*, 12 August 1954

GAINSBOROUGH 1956
R. Gainsborough, 'Frank Auerbach', *Art News & Review*, 7 January 1956

BONE 1956
Stephen Bone, *Manchester Guardian*, 11 January 1956

SYLVESTER 1956
David Sylvester, 'Young English Painting', *The Listener*, 12 January 1956

THOMPSON 1956
David Thompson, 'The Modish and the Odd', *The Times*, 13 January 1956

BERGER 1956
John Berger, 'Several Exhibitions', *New Statesman*, 14 January 1956

RUSSELL 1956
John Russell, 'Colourmen', *Sunday Times*, 15 January 1956

LESSORE 1956
Helen Lessore, 'Mr Auerbach's Paintings', letter to the Editor, *The Times*, 3 February 1956

SUTTON 1958
Keith Sutton, 'Remarkable Pictures', *Art News & Review*, 25 October 1958

BERGER 1959
John Berger, 'A Stick in the Dark', *New Statesman*, 28 November 1959

WALLIS 1959
Neville Wallis, *The Observer*, 29 November 1959

THOMPSON 1959
David Thompson, 'Mr Frank Auerbach's One-Man Show – Painter to be reckoned with now', *The Times*, 30 November 1959

SYLVESTER 1961
David Sylvester, 'Nameless Structures', *New Statesman*, 21 April 1961

WALLIS 1961
Neville Wallis, 'Two Inquiring Realists', *The Observer*, 23 April 1961

THOMPSON 1961
David Thompson, 'Uncompromising Painting – Mr Auerbach's Exhibition', *The Times*, 26 April 1961

SPENDER 1961
Stephen Spender, 'Mr. Auerbach's Latest Painting', letter to the Editor, *The Listener*, 27 April 1961

THOMPSON 1962
David Thompson, 'Mr Frank Auerbach's Paintings', *The Times*, 10 February 1962

BOWNESS 1962
Alan Bowness, 'At the Galleries', *The Observer*, 11 February 1962

ROSE 1962
J. Rose, 'Art', *Time & Tide*, 15 February 1962

SALVESEN 1962
C. Salvesen, 'Frank Auerbach', *The Arts Review*, 24 February 1962

SYLVESTER 1963
David Sylvester, 'British Painting Now', *The Sunday Times* colour magazine, 2 June 1963

RUSSELL 1963
John Russell, 'How will they be at forty?', *The Sunday Times*, 8 September 1963

FORGE 1963
Andrew Forge, 'Auerbach and Paolozzi', *New Statesman*, 13 September 1963

THOMPSON 1963
David Thompson, 'New Paintings by Auerbach', *The Times*, 16 September 1963

Selected Books, Exhibition Catalogues and Articles relating to Auerbach

(in alphabetical order)

ACKROYD 1994
Peter Ackroyd, 'Introduction', in *Frank Auerbach: Recent Works* (exhibition catalogue), Marlborough Gallery, New York, 1994

BRITISH COUNCIL 1995
From London: Bacon, Freud, Kossoff, Andrews, Auerbach, Kitaj (exhibition catalogue), The British Council, Scottish National Gallery of Modern Art (and international tour), 1995

COMPTON 1987
British Art in the 20th Century (exhibition catalogue), S. Compton (ed.), Royal Academy of Arts, London, 1987

FEAVER 1976
William Feaver, 'Introduction', in *Frank Auerbach, Paintings and Drawings 1954–1976* (exhibition catalogue), Marlborough Galerie, Zurich, 1976

FEAVER 2009
William Feaver, *Frank Auerbach*, New York (Rizzoli) 2009 (forthcoming)

FORGE 1968
Andrew Forge, 'Introduction', in *Helen Lessore and the Beaux Arts Gallery* (exhibition catalogue), Marlborough Fine Art, London, 1968.

GARLAKE 1998
Margaret Garlake, *New Art New World: British Art in Post-War Society*, New Haven and London (Yale), 1998

GOODING 1989
Mel Gooding, 'Introduction', in *Frank Auerbach: Recent Work* (exhibition catalogue), Rijksmuseum Vincent Van Gogh, Amsterdam, 1989

GOODING 1990
Mel Gooding, 'Introduction', in *Frank Auerbach: Recent Work* (exhibition catalogue), Marlborough Fine Art, London, 1990

HARRISON 2002
Transition: The Art Scene in the 1950s (exhibition catalogue), Martin Harrison (ed.), Barbican Art Gallery, London 2002

HUGHES 1990
Robert Hughes, *Frank Auerbach*, London (Thames & Hudson) 1990

HYMAN 2001
James Hyman, *The Battle for Realism: Figurative Art in Britain During the Cold War, 1945–1960*, New Haven and London (Yale) 2001

KOSSOFF 1978
Leon Kossoff, 'The Paintings of Frank Auerbach', in *Frank Auerbach* (exhibition catalogue), Hayward Gallery, London, 1978

LAMPERT 1986
Catherine Lampert, 'Introduction', in *Frank Auerbach* (exhibition catalogue), XLII Venice Biennale, 1986

LAMPERT 2001
Frank Auerbach, Paintings and Drawings 1954–2001 (exhibition catalogue), Catherine Lampert (ed.), Royal Academy of Arts, London, 2001

LESSORE 1986
Helen Lessore, *A Partial Testament*, London (Tate) 1986

MATHIS 1993
F. Mathis, *Frank Auerbachs Oeuvre : Untersuchungen zur Farbgestaltung im kontextuellen Umfeld* (published PhD thesis), Frankfurt am Main and New York (P. Lang), 1993

MORPHET 1984
The Hard-Won Image: Traditional Method and Subject in Recent British Art (exhibition catalogue), Richard Morphet (ed.), Tate, London, 1984

PODRO 1969
Michael Podro, 'Introduction', in *Frank Auerbach* (exhibition catalogue), Marlborough-Gerson Gallery, New York, 1969

SPENDER 1982
Stephen Spender, 'Introduction', in *Frank Auerbach: Recent Paintings and Drawings* (exhibition catalogue), Marlborough Gallery, New York, 1982

Selected Books relating to the Rebuilding of London

(in alphabetical order)

ACKROYD 2000
Peter Ackroyd, *London: The Biography*, London (Chatto & Windus) 2000

BARNETT 1986
Correlli Barnett, *The Audit of War: The Illusion and Reality of Britain as a Great Nation*, London (Macmillan) [1986] 1987

BULLOCK 2002
Nicholas Bullock, *Building the Post-war World: Modern Architecture and Reconstruction in Britain*, London and New York (Routledge) 2002

CHERRY 1988
G.E. Cherry, *Cities and Plans: The Shaping of Urban Britain in the Nineteenth and Twentieth Centuries*, London (Edward Arnold) [1988] 1993

ESHER 1981
Lionel Esher, *'A Broken Wave': The Rebuilding of England 1940–1980*, London (Pelican) 1981

FORSHAW AND ABERCROMBIE 1943
J.H. Forshaw and Patrick Abercrombie, *City of London Plan, Prepared for the London County Council*, London 1943

HOLDEN AND HOLFORD 1947
C.H. Holden and W.G. Holford, *The City of London: A Record of Destruction and Survival (Final Report of the Planning Consultants)* London (Town Planning Committee of the Corporation of London) [1947] 1951

HUMPHRIES AND TAYLOR 1986
S. Humphries and J. Taylor, *The Making of Modern London 1945–1985*, London 1986

JACKSON 1970
Anthony Jackson, *The Politics of Architecture. A History of Modern Architecture in Britain*, London (Architectural Press) 1970

RICHARDS 1942
J.M. Richards, *The Bombed Buildings of Britain: A Record of Architectural Casualties, 1940–41*, London (Architectural Press) 1942

ROSE 1985
Jack Rose, *The Dynamics of Urban Property Development*, London and New York (E. & F.N. Spon) 1985

SAINT 1987
Andrew Saint, *Towards a Social Architecture: The Role of School Building in Post-war England*, New Haven and London (Yale) 1987

SHEPPARD 1948
Richard Sheppard, *Building for the People*, London (George Allen & Unwin Ltd.) 1948

First published to accompany the exhibition

FRANK AUERBACH
London Building Sites 1952-62

at The Courtauld Gallery
Somerset House, London
16 October 2009 – 17 January 2010

The Courtauld Gallery is supported with funds from the Higher Education Funding Council for England (HEFCE)

HIGHER EDUCATION hefce FUNDING COUNCIL FOR ENGLAND

ISBN 9781903470947

British Library Cataloguing in Publication Data

A catalogue reference for this book is available from the British Library

Produced by Paul Holberton publishing
89 Borough High Street, London SE1 1NL
www.paul-holberton.net

Designed by Philip Lewis
Origination and printing by
e-graphic, Verona, Italy

FRONT COVER *Rebuilding the Empire Cinema, Leicester Square*, detail cat. 13
BACK COVER *Study for Shell Building Site: from the Festival Hall*, cat. 8 (iii)
TITLE PAGE *Shell Building Site: from the Festival Hall*, detail cat. 8
PAGES 10–11 *Shell Building Site: Workmen under Hungerford Bridge*, detail cat. 7

Photographic Credits

All works by Frank Auerbach
© Frank Auerbach

All images as credited in the captions with the exception of the following:

Frontispiece and fig. 37 courtesy of Beaux Arts Gallery Archive, Mr Henry Lessore; fig. 1 © The Lewinski Archive at Chatsworth; figs. 2, 6, 7, 13, 14, 26, 27, 29, 42, 43, 51 and cat. 1, 3, 4, 5, 6, 7 (i), 8 (i), 8 (ii), 8 (iii), 9 (i), 11 (i), 11 (ii), 11 (viii-xi), 12 (i), 13 courtesy of Marlborough Fine Art, London; fig. 4 © Estate of John Piper/Museum of London; fig. 5 The Bridgeman Art Library, © ADAGP, Paris, and DACS, London, 2009; figs. 8, 30, 31 © Leon Kossoff; fig. 10 RMN, © ADAGP, Paris, and DACS, London, 2009; fig. 11 © The Willem de Kooning Foundation, New York/ARS, NY and DACS, London, 2009; fig. 28 © Daniel Larson; fig. 32 © The Artist's Family; fig. 33 © Camera Press, London.